The Gene Age

ALSO BY EDWARD J. SYLVESTER

Target: Cancer

The Gene Age

GENETIC ENGINEERING
AND THE NEXT
INDUSTRIAL REVOLUTION

Edward J. Sylvester
and Lynn C. Klotz

REVISED EDITION

Charles Scribner's Sons · New York
Collier Macmillan Publishers · London

Charles Scribner's Sons
Macmillan Publishing Company
866 Third Avenue, New York, N.Y. 10022
Collier Macmillan Canada, Inc.

Library of Congress Cataloging-in-Publication Data
Sylvester, Edward J.
The gene age.
Includes bibliographies and index.
1. Genetic engineering. I. Klotz, Lynn C. II. Title.
TP248.6.S94 1987 660′.6 87-16539
ISBN 0-684-18819-8

10 9 8 7 6 5 4 3 2 1

Designed by Jack Meserole

PRINTED IN THE UNITED STATES OF AMERICA

TO OUR PARENTS,

EDWARD AND ELLEN SYLVESTER

AND

AARON AND GRACE KLOTZ

CONTENTS

PREFACE AND ACKNOWLEDGMENTS

Lynn Klotz and I initially set out to write a book that would bring into sharp focus the genetic engineering revolution going on around us, a revolution whose impact we believed would increase exponentially in years to come. We wrote *The Gene Age* in part to explain the science of genetic engineering, for the first time at a layman's level, much as writers and scientists worked in the 1950s to bring the stories of space travel, astronomy, and nuclear physics to the wide-ranging audiences whose lives were to be affected by those scientific revolutions. And we attempted an honest discussion of the risks and flaws of this developing industry in time for people to come to grips with them.

The result, in 1983, was the first such book on the subject. Lynn, whose career as a scientist kept him at the forefront of the new revolution in biology at the molecular level, brought to that effort his background in the biochemistry and molecular biology departments at Harvard and biochemical sciences department at Princeton and his then-new experiences as a founder of a genetic engineering firm. As a lifelong journalist, I was the interviewer of the dozens of leaders in the field who were Lynn's colleagues in science and business, and the chief writer.

Lynn and I felt from the start that if we could satisfy each other's demands in this book, we could produce something of value to all those with an interest in the developments in molecular biology within the last quarter-century. We believe that includes everyone, whether the interest lies in business, in science, or just in seeing what shape the future is taking. We hope and believe the book will interest those with a great deal of knowledge of the subject, yet be readily accessible to virtually any reader.

The revised edition represents a brand-new look at this exciting field that so far has exceeded our own forecasts in virtually every specific. Nearly everyone interviewed for the first edition was reinterviewed for the current effort, and dozens of additional sources were added to tell of the many developments in the few years since 1983. I believe the earlier material we've retained offers a depth of field and perspective for this new edition. For example, the importance of the still-modest successes in using interferon, Interleukin-2, and tumor necrosis factor to fight several forms of cancer is best understood against a historical background. In 1980, interferon was as mysterious as when it was discovered some thirty years earlier, valued at $10 billion a pound because the world quantity was measured in micrograms. Genetic engineering techniques made possible the mass production of interferon for use in clinical trials at dozens of hospitals, and the discovery of the latter two drugs.

Perhaps equally vital to understanding genetic engineering is the explosive growth of the industry that has sprung up in this decade and the mounting concern over its potential risks, which we deal with throughout the book as well as in chapters concentrated on these subjects.

Retained from the first edition with few changes needed are the three core chapters of the book on the science of genetic engineering. The reader with little scientific interest can skim these chapters, gleaning the key concepts needed to understand what's happening, referring to the index to return to basic definitions as required. On the other hand, those wanting to develop a solid background in the scientific concepts involved in gene splicing can read the three science chapters in sequence, as a primer, and then pick and choose among the remaining offerings on the impact of this science in various areas of life.

We designed the science sections to deliver the essentials of molecular biology *as it relates to genetic engineering;* to keep readers from being lost in a whole new universe of particles, functions, and terminology, we have trimmed the science to the very barest skeleton, yet have tried to leave out nothing essential to an understanding of what scientists mean when they speak of splicing genes, achieving protein expression, or that favorite new word, *cloning* things. We coined the term *genefacture* to encompass the industrial processes harnessing the power of genetically altered microbes, but other than that the terms are those already in use by genetic engineers. Think of *The Gene Age* as a guidebook to an unknown country, which attempts to teach a new language as well as to convey the complexity of a culture.

Deeper background: Lynn and I have been talking about the potential wonders of genetic engineering for nearly twenty-five years now, since our roommate years at Princeton University in the mid-sixties. We burned many a midnight lamp talking about the developments he was witnessing as a student in Jacques Fresco's laboratory, as the properties of the life-control molecule DNA were being discovered. Our wildest speculations were of the kinds of discoveries this book is about, but neither of us foresaw their occurrence in our lifetime. In fact, much of the science that we discuss at layman level here was not known to leading molecular biologists at that time—witness Sir Francis Crick's comments on that period in Chapter 3.

Most startling to Lynn: despite his interest in the subject, he never suspected he would be part of the scientific revolution. And perhaps most startling to me was how his comments on the smallness of this field would be brought home. Lynn moved into our dormitory group to replace another Fresco protégé who had been so successful in his undergraduate studies that he had finished early—almost unheard of at Princeton then—and had moved on to graduate work at Northwestern University. Two decades later, Thomas Wagner would gain fame at the University of Ohio for his work on "three parent" cattle and transgenic livestock. As we spoke in the fall of 1986, for the first time since undergraduate days, Wagner said of the "genies" (see Chapter 3), "We're really such an unusual group to be so involved in an industrial enterprise. None of us then had any orientation toward business or finance, none of us thought in terms of what practical application might be made of what we were doing; we'd have laughed off anyone who even suggested such a thing."

A fitting capstone to our own interviews and observations along those lines, followed by an interesting speculative question from Wagner: "I wonder what will happen as the next generation comes along. What kind of ethical behavior will result when you have scientists who went into the field knowing what sort of potential for profit there is?"

Since most of *The Gene Age* is based on interviews attributed directly in the text, any material taken from published information is cited in notes listed by chapter at the end of each chapter. These notes may also serve as a guide to further reading, but they do not contain supplementary information—that is, the reader need not flip to the notes for explanations in order to follow the discussion.

Most of those who gave so generously of their time and energies to help us convey genetic engineering to the reader appear within the book, although generally their contributions went far beyond their relatively brief quotes, be they Ronald Cape, founder and chairman of Cetus, or genetic engineering critic Sheldon Krimsky of Tufts, both of whom provided a great deal of written background material. Similarly, we would like to thank William Gartland, executive director of the Recombinant Advisory Committee in the National Institutes of Health; Morton Collins, whose venture capital business is headquartered in Princeton, New Jersey; Tom Roberts of Dana Farber Cancer Institute; Lewis Cantley of Tufts University Medical School; Vicki Sato, director of Cell Biology and Immunology at Biogen in Cambridge, Massachusetts; and John Hunt, chairman of BioTechnica International, Cambridge. We would especially like to thank David Glass, director of patents and regulatory research of BioTechnica, for providing a great deal of information and for reading the manuscript.

Neither this edition nor the first would have been possible without the talent, energy, and support of our book editor, Elizabeth Rapoport of Charles Scribner's Sons. Finally, as in all my work, I was immeasurably helped by my wife, Ginny.

Drawings for *The Gene Age* are by Jenny Lodge and Suzanne Lawson. For the revised edition, May Chaney of Biocommunications, Phoenix, supplied Figure 3A.

ED SYLVESTER

Tempe, Arizona
March 1987

The Gene Age

INDUSTRY COMES TO LIFE

1

Tiger! Tiger!
Burning bright
In the forests of the night,
What immortal hand or eye
Could frame thy fearful symmetry?

MAN invents tools and the tools change man. So goes the adage. But the change is not instantaneous, and those who watch new tools work magic on their inventors observe three stages. First, people use a new technology to accomplish something they've been doing all along, only now can do better—faster, more easily, perhaps ultimately in such mass quantity that what had been a luxury of the privileged becomes a popular commodity. Then pioneers move on to put the discovery to new uses, to accomplish goals once thought impossible or inconceivable. Finally—surely the most interesting and hardest-to-forecast stage—the new tools alter the very shape of the society in which their inventors live, work, and invent. Mankind is changed, the context and the relationships of human society altered deeply and permanently.

The industrial revolution swept across Europe and then the New World for two hundred years, transforming quiet green countrysides dotted with millponds into industrial milltowns—the mill central to both, the idea of a mill radically changed. Mills were water-powered, at first by falling rivers, soon by steam in industrial boilers using enormous energy to create high temperature and high pressure. Then ever-increasing quantities of energy powered hearths to create virtually every product of modern life. Some say William Blake's tiger, burning bright and terrible, was his vision of the industrial revolution at its beginning. The hallmark of that revolution was scale: We learned to use natural resources in unprecedented kind and quantity, transformed them by using still other resources to produce higher temperatures and pressures, spread the results farther, and surely transformed society on a scale never before approached.

Consider those three stages of transformation as they relate to just one creature of the industrial revolution, the internal combustion engine. Invented only a century ago, this engine entered the nineteenth century innocently, almost comically, as a dubious replacement for the horse to pull carriage or plow. But in this century it made possible such diverse inventions as the airplane and the submarine, which, without engines, would have remained glider and diving bell, intellectual curiosities, hardly reshapers of society. Finally, the engine has led us to a global society, of distant countries suddenly close, often too close for comfort, of global warfare, and in our own country a society of suburban living and shopping and freeway commuting—a pattern of life completly unheard of a century ago. The engineer–mathematician–venture capitalist who pointed out those stages noted as a measure of their magnitude that his own field, electronic data processing, is only at the end of stage two. For hints of the last stage, he suggests, look to satellite communications completing global interdependence, and to space travel, both children of the computer age as surely as are nuclear weapons.

Genetic engineering promises a revolution more far-reaching than that wrought by the computer, one that may bring to a close the industrial revolution itself. Now, at its birth, it is essential that we confront its potential. Few scientific breakthroughs have been the subject of more early overstatement. Yet if we look into the future, around the corner of a new century and millenium barely a dozen years away, it is impossible to exaggerate the potential of genetic engineering for good, and if misused, for evil.

In the industrial revolution we harnessed nature's brute forces to produce the hardware of modern economy. But its hearths demanded coal stripped from the earth, electric power from dammed rivers, more and more of a dwindling supply of oil. Along with shortages we got an abundance of noxious wastes and an environment fouled by smoke.

Genetic engineering is the hearth of the next industrial revolution, and in that hearth industries are already forging new versions of current products and the first of an array of products never seen before—produced without the industrial hangover. Energy-intensive furnaces and high-pressure boilers are giving way to sterilized vats in which chemicals can be grown as by-products of bacterial colonies, or formed through the intercession of the enzymes that regulate life. This "genefacture" is carried out at ordinary air pressure and at the temperature of a fine spring day.

Genefacture—genetic industrial manufacture—should be a much cleaner industry than we have known and, properly regulated, a safe industry. This is an important theme that recurs throughout the unfolding story of *The Gene Age.* In fact, instead of causing pollution, microbes are already being developed that may some day devour garbage and give off valuable methane gas as a bonus, and they may enable us to clean up toxic pollutants. The early versions of such microbes already exist. An oil-eating "bug" got America's first-ever patent for a novel life form. Its inventor says it does not operate on a scale large enough to have practical use in fighting oil spills, but an eventual descendant might. Since then, that scientist has developed a microbe he says will devour dioxin, the controversial suspected carcinogen in Agent Orange that contaminates acres of land at U.S military bases. Other firms are testing bacteria that will devour a wide rage of industrial soil pollutants and pesticides (see Chapter 10).

This is the era of the high-tech entrepreneur, as we are often reminded, and genetic engineering is as much the creation of the entrepreneur as of the scientist. It is a fusion of pure science and business, of laboratory and market, to a degree unmatched. Never before has such a hybrid offered so much in so short a time or attracted such a clamorous response from the investment community, the public, and scientists themselves. An industry that virtually sprang to life in 1980 now has drawn an investment of more than $2.5 billion—much of it before it could offer a single product or a penny in revenue, and that in a country whose business focus is notoriously short-range. How? Some of it owes to the idea of venture capital (see Chapter 5), an American enterprise just catching on in Europe.

But this is only one side of genetic engineering, and if it indicates its sweep, its breadth, we have said nothing of the depth of its potential influence on us. As we'll see, the tools of genetic engineering arose from basic discoveries about the growth and development of living things, and in genetic engineering we are using those tools to redirect this growth and development. Only five years ago scientists spoke of "gene therapy" as many decades away. That one could tamper with the human genetic inheritance was and is an idea with overwhelming moral implications; but to do so in order to wipe out dreaded inherited diseases is generally regarded as benign. Nevertheless, the ability to change genes in a fetus or a living person was regarded as one for the next century. Now, a pioneer in genetic engineering techniques says that within a decade he believes it will be possible to alter the genes of living people if their malfunction is life-threatening.

Recently, human growth hormone (HGH) went on the market. Scientists genetically engineered HGH to eradicate dwarfism; but now there are reports that some physicians are prescribing it for their children, to ensure that they grow tall in a society that respects height. How tall is the right height? Within this decade we may find ourselves debating that question. But, the adage holding, nowhere is the effect of these new tools more profound than within the basic science that created them. Molecular biologists are coming to an understanding of how cancer develops, how mutation occurs, even closer to understanding how a single fertilized egg becomes a human.

A decade ago the pathways in the development of cancer were little better known than they were a century before that. Thanks to the tools of recombinant DNA, molecular biologists now have zeroed in on genes they believe play a key role in cancer development, and in late 1986 the first gene was positively identified that plays a definitive role in one cancer (see Chapter 10).

Most attention naturally has focused on powerful new drugs, vaccines, and diagnostic techniques emerging from biotechnology, both because they are already beginning to reach the public and because any discovery that affects our health is important, personal, and for those reasons, dramatic. Most leading scientists now believe we can win the war on cancer, and the end of this story brings us to the pursuit of cancer's causes and treatments with such new drugs as Interleukin-2, interferons, tumor necrosis factor, and monoclonal antibodies. Genetic engineers have found among the trace molecules of the body one that appears to aid heart-attack victims in recovering, a protein called Tissue Plasminogen Activator (TPA). These terms will soon be clear.

Five years ago, a group of scientists led by Dr. Thomas E. Wagner at the University of Ohio got growth-inducing genes from rats to function in mice—the first time that genes from one mammalian species had been made to function in another. By last year, Wagner's group had moved all the way to creating a "transgenic pig," that is, one that had the inherited genes of its parents plus new genes spliced into its chromosomes. Three or four scientific teams here and in Australia were on their way to getting recombinant animals "on the ground," to follow the current lingo of the transgenic barnyard. To the casual observer such achievements often seem the stuff of humor: Who wants to build a bigger mouse? What's in a better pig? By the time this story is done, the wide-ranging impact of those events should be much clearer. The scientists' immediate goal is to make food animals develop more meat and less fat and consume feed more efficiently. This, in essence, means money savings to farmers; it could mean more meat and milk from a given small amount of feed.

But Wagner sees the promise of such genetic engineering in domestic animals as far beyond marginal production increases. He foresees the marginally profitable (or unprofitable) family dairy farm as a producer of drugs and specialty chemicals (see Chapter 10), a notion that several other scientists began considering several years ago for crop farmers.

And if such gene manipulation can be accomplished in pigs, then how soon could we correct devastating genetic errors in humans that cause such diseases as the blood and bone disorder beta thalassemia, or sickle-cell anemia, or Tay-Sachs disease?

Not surprisingly, this field of great promise is a battlefield: among executives in hot pursuit of profitable products (see Chapter 5); among scientists in universities concerned with the effects such practical-minded pursuits will have on the academic search for knowledge; among members of Congress concerned with the use of public money (see Chapter 7); among clergy and others who fear we have again advanced intellectually without sufficiently considering ethical implications or unforeseen side effects, such as the possibility of releasing a hazardous new microbe into the environment (see Chapter 8). These important debates unfold in the context of a science story, a tale told in a submicroscopic world whose actors move far beyond the reach of the eye. *The Gene Age* is a story about the meeting between the cutting edge of science and the "real world" in which we live and work, bear and raise our children, and seek permanent solutions to our gravest problems.

This story's "central figure," the key to genetic engineering and the secret of all heredity, is a long-chain chemical molecule called deoxyribonucleic acid—DNA. This molecule, found within every living cell, controls the development and function of all life on earth. Whether the cell is one of billions making up a complex human or the single cell of a bacterium, it contains at least one molecule of DNA. DNA functions much like an architectural blueprint, instructing the cell to produce the proteins essential to its survival. When scientists or doctors speak of "genes," they are referring to something which as a physical entity is *nothing more than a region along the DNA molecule.*

Just over ten years ago at Stanford University, Dr. Paul Berg discovered a way to splice genes (that is, DNA regions or fragments) from one kind of virus into the DNA of another kind of virus—he "recombined" the genetic information by recombining DNA fragments. That was the first *recombinant DNA* experiment. His plans to go one step further and insert viral genes into the DNA of the bacterium *Escherichia coli (E. coli* for short) began the long controversy over potential risks to humans in such experiments, as we'll see.

The controversy over recombinant DNA began within the biological science community but spread much farther, and although many of the scientists' initial questions have now been answered, others remain. The major fear was based on the fact that *E. coli* is one of many types of bacteria that dwell harmlessly in the human intestines: What if an experimental strain were turned into a deadly pathogen and escaped from the lab? Would we face a plague unmatched in history? In this debate Berg himself favored extreme caution. Guidelines were set up by the federal government, and eventually some of the initial caution was seen as overreaction.

The way has been opened for genetic engineering. Scientists immediately realized that if you could instruct a rapidly reproducing *E. coli* to make a virus protein along with the proteins it normally manufactures for its own livelihood, then you ought to be able to instruct it to make a valuable protein such as human insulin. Or human growth hormone. Or interferon. Engineering genes into *E. coli,* yeasts, and other organisms so that they would instruct these hosts to make proteins or other chemicals of great human value has been the consuming effort in the years since. Making the products we discuss in this chapter involves the basic process of inserting the foreign DNA carrying instructions for a valuable enzyme, hormone, or other protein into the DNA of some other organism, so that the host organism makes the desired protein at the same time it makes its own. That's our definition of genetic engineering.

Once you've got the redesigned bacterium, there's nothing left to do but grow a whole colony from that single parent in the process called *cloning,* a procedure that exploits the principle that bacteria reproduce by division of a single "parent" cell into identical "daughter" cells.

The keys to high performance in genetic engineering can be stated simply: Rebuild a bacterium's DNA so it makes the right protein in the largest possible quantities, and grow a large colony from that single parent. As rapidly as that can be done on an industrial rather than laboratory scale, the promise of genetic engineering will be realized.

But what a realization that should be. The U.S. Office of Technology Assessment (OTA) predicted in mid-1981[1] that genetic engineering could displace standard manufacture of products valued at $27 billion over the following twenty years. That's the approximate size of the American organic chemical industry now. Furthermore, the man who provided much of the supporting data for that forecast now believes he erred on the conservative side. Dr. Leslie Glick, president of Genex Corporation, which is based in the Washington area and is one of the earliest gene-splicing companies, was a key consultant for the OTA study. Based on "good, hard data," Glick reassessed his calculations two years later and came up with a $40 billion estimate for the value of genetically engineered products by the turn of the century.

A look inside Glick's calculations now reveals some of the vagaries inherent in such forecasts. For example, the $40 billion was broken down into $14 billion worth of existing products, half of those produced from petrochemicals, and $26 billion in brand-new products on the market. Glick now believes his forecast to be remarkably on target despite wild changes in the parameters he used to calcuate it, such as the price of oil. "No one in 1981 expected the price of oil to drop at all, ever, let alone to plummet. It was then around thirty dollars a barrel; now [in late 1986] it's fifteen dollars. That has certainly put a damper on research and development into petroleum replacements."

Chief among these, of course, are ethyl alcohol as a fuel substitute and methane for fuel, cooking, and heating, both early targets for large-scale genetic engineering technology. But Glick is quick to point out, "There's only so much oil on the planet. We really ought to be looking for a replacement before we need it; at least we'd be a lot farther along when the time comes."

A recent analysis Glick did for the Industrial Biotechnology Association, an industry trade group, arrives at the same $40 billion forecast, albeit via somewhat different routes. Of major importance is the unexpected speed with which promising new pharmaceuticals have been developed and, thanks to fast FDA approval, brought to market. A good example of how such developments change valuations and make forecasting a guessing game can be seen in the story of interferon. Discovered twenty years ago, this tiny, elusive molecule exists in the human body in such small quantities that nothing was known of its structure and function as the 1980s began. It was thought to somehow "interfere" with viruses attacking the body, hence the name.

In 1981, a Harvard medical research newsletter set the price of interferon at $10 billion a pound—but of course, there wasn't a pound of it in all the world's laboratories, and that was the point. Dr. Orrie Friedman, president of Collaborative Research in Waltham, Massachusetts, noted then that once the new gene-splicing companies got sizable batches out, the price would plummet—it was still unproved as a drug, and it would no longer be rare. His words were prophetic. By 1983, any qualified researcher in the world could get a sizeable sample of several different types of genetically engineered interferon free for the asking. Its makers hoped someone would find a use for it—strictly speaking it was worthless, in plentiful supply yet unproved. Suddenly, in 1985, all that changed. Interferon achieved remarkable results in chemotherapy against hairy-cell leukemia, one of the most fatal cancers and one most resistant to other forms of chemotherapy. In June 1986, the FDA approved alpha interferon for general use as an anticancer drug; manufacturer Hoffman La Roche, which holds patents on the drug, stands to make substantial profits from its sale. Still other developments, to be discussed shortly, suggest that interferon may be an early "boom" development of genetic engineering, although the first multimillion dollar product may well be Tissue-specific Plasminogen Activator, for treatment of heart attack patients.

We can map out the scope of the genetic revolution on three fronts, remembering that the largest impact of all is on our understanding of ourselves as living beings, and that "doesn't compute" in the usual sense. But we can measure changes now and forecast them for tomorrow in medicine, industry, and agriculture.

Medicine

Vaccines You sneeze. The flu is coming on. That means viruses have gotten into your body and attached themselves to a few cells and worked their way inside. Viruses may be the world's simplest "organisms," if organisms they are: although they come in a multitude of forms, they are nothing more than DNA or RNA (ribonucleic acid, a genetic molecule very similar to DNA) wrapped in a protein coat. But watch them in action. By some type of chemical trigger, the viruses insert their DNA into your cells. Now, in nature's own version of genetic engineering, the viral DNA takes over the reproductive mechanism of your own cells, forcing it to reproduce viral DNA along with your own. Hosts of new viruses are now made in your cells, usually budding out from the cell surfaces, then spreading to carry the invasion. You don't feel a bit well. But you're not whipped yet.

Viruses have certain surface markings as easily readable as an invading jet's are to an air observer. Antibody cells constantly swimming in your blood recognize these invaders, or antigens, and attach to them. Antibodies are very specific: each type recognizes only one type of invader. This initial "interception" of the invaders also appropriately triggers a general mobilization—the production of vastly more antibody cells, which secrete huge amounts of antibodies into the bloodstream. They also attach to invaders and literally sink them out of the blood. The war is on. Eventually the antibodies win: you get better.

This or a similar scenario unfolds at the onset of all viral infections, and the antibody interception marks the starting point for cure and prevention. In order to vaccinate against a particular flu, for example, doctors take a batch of the virus causing that flu, kill it, and inject it into the bloodstream. If all goes well, the antibodies recognize the surfaces of the *dead* virus and sound the general alarm, producing more antibodies. Thus the antibodies are massed in advance to repel future invasions of the same flu virus before it gets there. Polio vaccine works in much the same way, and it protects for life.

But there is an inherent hazard in all so-called killed-virus vaccines. If the virus by accident is not completely deactivated, patients may get the disease instead of becoming immunized against it, an occurrence frequent enough to make many such vaccines unpopular for general use around the world. Further, there are some dead viruses to which the body's immune system for some reason will not respond, so antibodies don't multiply in preparation for attack. Killed-virus vaccine serves no purpose against those diseases.

Genetically engineered vaccines could solve these problems, because they are not composed of killed viruses but only of virus *parts*. Ultimately, perhaps, the parts may consist of only those surface markings that trigger antibody response. Such vaccines would never contain the whole virus, so they would be completely safe.

The first genefactured human vaccine approved by the FDA for general use was one against hepatitis, and a great deal of the promise and problems in genetic engineering are reflected in its story. Hepatitis is a serious viral disease of the liver, frequently leaving its victims with permanent liver damage if they survive, and at high risk for getting hepatoma, one of the deadliest forms of cancer. Although relatively rare in the United States, hepatitis is endemic in the Third World, and the liver cancer associated with it also has a high incidence in poor countries.

The first hepatitis vaccine was purified from huge quantities of human blood, donated by those who had had the disease and so had antibodies to the virus in their bloodstreams. But those getting the vaccine were now at risk for acquired immune deficiency syndrome, or AIDS, a too-frequent passenger in large quantities of donated blood. Finally, William Rutter of the University of California at San Francisco created a hepatitis vaccine by recombinant DNA techniques. Because this vaccine was made in microbes and made no use of human blood, it is one that leaves the recipient at no risk of getting either hepatitis from a killed virus or AIDS from impure blood. The vaccine is now being marketed, but a court fight is underway between rival biotechnology firms over who has patent rights, which might allow one firm exclusive rights to market the product—another sign of the times (see Chapter 6).

Finally, an antimalarial vaccine has been created and human clinical trials are to begin soon. Malaria is probably the world's leading disease, with 800 million sufferers now in the world. A host of wasting diseases with lesser known names trails closely behind. In 1986, Rockefeller Foundation officials announced a $300 million campaign against these killers—a campaign they said was largely made possible by advances in biotechnology.

Hormones and other enzymes Both simply forms of protein, they form the largest class of genetically engineered pharmaceuticals to be produced and marketed via genetic engineering, the best known being insulin. Originally expected to be through testing stages and in the hands of diabetics in late 1983 in the United States, Genentech's "Humulin" passed its FDA requirements and went to market in late 1982.

The clamor over genefactured insulin confused some people, who had expected the genetically engineered product to be cheaper. It is not, but the significance of Humulin was that it is very pure and, more important, it is human insulin. Until 1982, diabetics had to take insulin gathered from the pancreases of slaughtered cows and pigs. Both animals' insulin is remarkably close to human, but not identical. Researchers suspect that allergic reactions among some diabetics were caused by these small differences.

Several firms are now selling human growth hormone (HGH). Not only should this prove a cure for dwarfism, but scientists believe that it or related growth hormones will prove major aids in healing burns and other wounds.

Researchers have genetically engineered such proteins as factor VIII, lack of which causes hemophilia, and another whole group of hormones called endorphins and enkephalins that have stirred excitement among medical researchers. The latter two are produced in the brain and are thought to be natural painkillers. In fact, these or similar hormones are thought to be responsible for "runner's high," the feeling of well-being experienced by long-distance runners and joggers. If so, scientists might eventually produce powerful yet nonaddictive painkillers and anesthetics by recombinant DNA means.

Interferon No little-understood protein has caused the excitement in research or attracted the publicity of interferon. First isolated about twenty-five years ago, interferon is a natural antiviral agent inside the cell, as antibodies are in the blood fluids outside the cell. There are actually nineteen or so similar proteins known as interferons. Scientists have long hoped that one or more would prove to be the long-awaited "magic bullet" for all viral disease, killing viruses with the efficiency that penicillin and other antibiotics wipe out bacterial infection.

That is still a pipe dream, but interferon is proving a valuable weapon in the anticancer arsenal, joining interleukin-2 and tumor necrosis factor, both discoveries of the Gene Age, in attacking cancerous cells. Alpha interferon was approved for general use as an anticancer drug in June 1986, but further testing must be conducted before it can be prescribed for other uses—and other uses appear just around the corner. Biogen, of Cambridge, Massachusetts, has applied for FDA permission to use its alpha interferon as a cold preventative. Initial experiments with interferon against colds had suggested it was not effective or caused such side effects as nasal ulceration—a cure worse than the disease. A Biogen scientist, however, says these problems have been overcome by properly regulating dose and formulation.

How interferons work still is not completely understood. But they apparently cut the efficiency of certain viruses' ability to make vital proteins and even help determine what proteins can be made for the virus, interfering with takeover of the invaded cell. One form, gamma interferon, is also a potent stimulator of the immune defense system.

Like interferon, tumor necrosis factor (TNF) and interleukin-2 (IL-2) are proteins produced naturally, but produced in such tiny quantities that until the advent of genetic engineering, they were too rare for experimentation. Now both are being used in clinical trials and with great success—although it is too early to speculate whether they will have widespread, permanent power to eliminate cancer. Cetus Corporation, one of the larger gene-splicing companies, and Roche Pharmaceuticals reportedly are preparing a court fight over royalties from IL-2. Meanwhile, Cetus-produced TNF is undergoing clinical trials at Fox Chase Cancer Institute in suburban Philadelphia. Of all the new anticancer weapons, TNF seems constitutionally most likely to succeed—but for one major problem. It is, as the name implies, a protein that promotes the death of tumor cells; however, it is highly toxic to other cells as well. Cetus scientist Mike Kriegler, formerly of Fox Chase, believes he can solve that problem.

Some of these trace proteins appear to have widespread uses. interferon and interleukin-2, for example, have shown extremely promising results against arthritis—a disease, crippling in its severest form, that results from the immune system's attacking certain of the body's own cells.

It is impossible to list all the naturally produced trace proteins that have been manufactured in quantity for the first time in history through genetic engineering, now being studied in laboratories around the world for their efficacy in medicine. Although what the body produces in tiny amounts can have serious side effects in quantity, the fact that these drugs are the body's own suggests that their eventual use will be less problematic than use of non-human compounds. Two such recently discovered proteins must be mentioned, however, because one, Tissue-specific Plasminogen Activator (TPA), produced immediate, exciting results, and the second, colony stimulating factor (CSF-1), was forecast as a long-range $1 billion product in the war against cancer.

TPA, made by Genentech and in clinical trials, has so far been highly successful in preventing blood clots in heart-attack patients. Doctors now use streptokinase, a bacterial product like the antibiotics, for this purpose. Problem: streptokinase is "untargeted"; in preventing clotting it can cause widespread bleeding. TPA has a binding site for a protein only produced as blood is clotting, so it will not work in other parts of the body.

CSF initially should be useful in safeguarding cancer patients who have had their bone marrow knocked out by chemotherapy. Chemotherapy patients are always at high risk of infection because their white blood cells and often their bone marrow have been destroyed or seriously impaired. Apparently, CSF marshals a range of other immune system agents. While bone marrow and white cells are rebuilding, CSF treatments may ward off disease. The CSF worked well in animal trials with monkeys, whose immune systems are most like humans'.

Monoclonal antibodies We described how antibodies are

the major element in the immune system's army of defenses against invaders, and how each antibody homes in on one and only one recognized site on the attacker. The human immune system therefore sends out multitudes of different antibodies against a multitude of different sites on any given invader.

For some time scientists had tried unsuccessfuly to isolate and raise colonies of *single* antibodies—antibodies that would recognize one site on one attacker. In 1975, two scientists did it. They fused myeloma cancer cells (which multiply extraordinarily fast) with antibody cells. Then they spread the resulting colony so thin that all the cells were widely separated—widely enough that any single cell could be grown into a whole colony separate from the others. This is the process of cloning, to be described in detail later. In this way, the scientists got whole batches of the same (monoclonal) antibody.

A serum of such single antibodies would flow quickly to one and only one site in the body, so scientists hope that monoclonal antibodies can be used to carry drugs, antibiotics, or even cell-killing radiation to specific sites of infection or tumor growth, leaving healthy tissue unaffected. That could eliminate the side effects of much existing radiation and chemotherapy. The long-term promise for such therapy remains high, but scientists have run into unexpected problems in getting these antibodies to carry drugs to disease sites. Various types of monoclonal antibodies are in clinical trials around the world, but none has yet proved itself a sure cure. The first to be used for a non-cancer application has just been approved for trial by the FDA. Called OKT-3, it shows promise in preventing rejection of transplanted organs. Like TPA, OKT-3 replaces a bacterially derived drug, cyclosporin, which sometimes has toxic side effects.

Diagnostics are a major area in which monoclonal antibodies should quickly replace existing techniques. Because they recognize specific antigens, these antibodies could be used in test kits for pregnancy, cancer, and virtually all infectious diseases. The kits would be inexpensive, safe, and far more accurate than many now available.

Precision diagnosis We are prone to ills carried by viruses, bacteria, fungi, and myriad other organisms, and diagnosing which one is responsible for a given disease is the key to its cure. Until now, such diagnosis for some diseases has been extremely difficult;

even when tests are available, they may give inconclusive results, be expensive to administer, or take a long time to complete. But every kind of organism has its own genetic "identity." It is *what* it is because of the very particular array of its genes along its DNA strands. A virus such as HTLV-III, the cause of AIDS, carries out its devastation through the workings of its particular DNA sequences, as surely as we are able to digest food and ward off most infection because of human DNA sequences.

Now genetic engineers are learning to identify segments of DNA representing the genetic signatures of a growing number of disease-causing microorganisms. These segments can then be used like magnets to probe for the existence of the same segment in a diseased patient. Unlike some other diagnostic probes, DNA probes are extremely sensitive to the identity of their targeted pathogens. This magnetlike effect occurs through the principle of "complementarity," to be explained in Chapter 2.

Consider traditional diagnostic tests. Generally, to identify what pathogen has the patient "down and out," a laboratory must grow up a colony of the culprit microorganism, a colony big enough for large-scale chemical effects to show. Naturally, the organisms must remain alive, and that can be difficult to accomplish in the lab. DNA probes could allow identification while the patient waits in the doctor's office, since there would be no waiting for a colony to grow. More important, they would allow the doctor to begin treatment immediately, before the disease progresses.

Now consider AIDS diagnosis, currently not done with DNA probes. Tests are so inconclusive that they indicate only whether a patient has been exposed to the HTLV-III virus, not whether in fact he or she is infected. DNA probes, which should be available soon, will prove *infection* or lack of it.

Monoclonal antibody tests work a little differently, although in both cases the probe is attracted to the pathogen. Antibodies fit lock-and-key fashion to very specific invaders. A colony of a single type of such antibody seeks out its own pathogen within the tested patient's bloodstream. Monoclonal antibody kits have been sold since the early 1980s, and tests for an ever-growing list of infectious agents are coming out daily. DNA probe tests are just beginning to appear. Such diagnostic tools may lack the dramatic impact of interferons and other potential anticancer drugs, but they may be responsible for real advances in disease care; and in the short term, they may be responsible for the survival of fledgling biotechnology companies by providing a much-needed revenue stream.

DNA probe diagnostic tests have now been developed for Legionnaire's Disease. Look for similar tests over the coming year for AIDS, hepatitis, herpes, and periodontal disease, among others.

Precision diagnosis II: Inherited traits For years it has been known that muscular dystrophy, hemophilia, sickle-cell anemia, Huntington's chorea, Tay-Sachs disease, and a host of others are directly inherited. All our inheritance comes in the form of DNA molecules comprising all the genes, the DNA in turn packed into the forms we call chromosomes; therefore, if we say a disease is inherited we mean it is reflected in the DNA of at least one parent, as well as the DNA of the child. Traditionally, genetic counseling has been imprecise, basing its probability forecast on family background, but except in rare cases, it has not been knowable with certainty whether a particular fetus in the womb has a given disease. Now recombinant DNA technology is allowing researchers to identify particular sequences of DNA that either are responsible for the disease or that always accompany the genes responsible—the latter known as genetic markers. Samples of fetal DNA can be obtained relatively safely, and in many cases the presence of the disease in the baby can be known with near-certainty.

DNA probes are already in use for Huntington's chorea, sickle-cell anemia, phenylketonuria, and hemophilia, among others, and the list is growing rapidly. More dramatically, researchers recently located and described the gene responsible for the major form of muscular dystrophy, leading to a hope that some day this wasting muscular disease may be eradicated.

But what about those genetic diseases not directly inherited, diseases that tend to run in families and, in many cases, develop only in later life? This long list includes diabetes, asthma, heart disease, and some forms of cancer; even alcoholism and such psychological conditions as depression now are believed to have strong genetic components. An intensive effort is underway to develop probes for susceptibility to these ills, but here troublesome ethical questions often emerge. If early diagnosis allows one to prevent the disease or reduce its effects, well and good. But what if no countermeasures are possible? In many cases, it may be better not to know what you're going to get in ten years. Considering the course of some cancers, on the other hand, even though early awareness might not directly lead to prevention or cure, monitoring may catch a tumor at an early stage, before it spreads, and therefore prove worth the worrying.

Fermentation Because the discoveries of genetic engineering are so recent, the aura of newness tends to extend to everything involved in biotechnology, but much of enzyme manufacture and other fermentation is ancient technology. Alcohol was produced by the Sumerians and Babylonians before 6000 B.C. Egyptians were using brewer's yeast to leaven bread in 4000 B.C. Production of purer alcohol through distillation of fermented grain was common in most parts of the world by the Middle Ages.

Louis Pasteur, identified by most of us with his work on rabies vaccine and "pasteurization" of milk, did much of his research to aid the French wine and beer industries. His studies of yeast revealed that this organism ordinarily respirates, like all other living things, using oxygen to burn food and giving off carbon dioxide as waste. But unlike most living things, Pasteur found, yeast can live without air by shifting metabolic gears into an entirely different way of deriving energy from food—anerobic (without oxygen) fermentation. Though not nearly as efficient an energy producer as respiration, this airless fermentation gives yeasts an edge in surviving, because it is carried out in an airless environment that would kill most organisms. The major waste product of fermentation: ethyl alcohol.

It might seem that alcohol production is the only ancient use of fermentation technology and therefore an anomalous example, but not so. Writing in a special *Scientific American*[2] issue on industrial microbiology in September 1981, Douglas Eveleigh noted that bulk fermentation is "not just a new field of entrepreneurial activity; it is a well-established factor in the world economy, responsible for a current annual production valued at tens of billions of dollars." Antibiotics certainly constitute a major product. The four most important groups of antibiotics—penicillins, cephalosporins, tetracyclines, and erythromycin—have bulk sales of more than $4 billion. They are all produced by bulk fermentation.

These drugs are natural metabolic products of the microbes from which they sometimes derive their names (penicillin, for example, is derived from *penicillium* mold). We consider vitamins "primary" metabolites because organisms create them as an essential part of their life-support processes; they are equally vital to humans, but we cannot make them and must get them from food. Antibiotics and other chemicals such as alcohol are "secondary" metabolites. Microorganisms make them either as waste products or—in the case of the "wonder drugs"—possibly to attack other bacteria that attempt to invade their living grounds.

Forcing such bacteria to mutate so they "overproduce" their natural antibiotic in greater and greater yields has been a practice in the pharmaceutical business for decades. For example, after Alexander Fleming discovered the antibiotic properties of penicillin in 1928, it took a dozen years for scientists to put the discovery to work and to force production levels needed for practical use of the drug. In fact, it was an American scientist working for the Department of Agriculture who developed the bulk fermentation process used in the first commercial production. Now, companies can get 10,000 times that first production level.

The U.S. Office of Technology Assessment (OTA) foresees major markets for genetically engineered, bulk-produced pharmaceuticals, including vitamins and antibiotics, within ten years. Pound for pound, these drugs don't earn as much for their makers, and they require high scale-up in capacity of fermentation vats and other equipment, but the markets for them are correspondingly huge. Vitamins alone represent a $500-million-a-year industry, and the OTA forecasts that genetically engineered organisms could soon produce them economically enough to displace current methods. Similarly, sources in the field predict antibiotic production could be aided by genetic engineering in two to four years, marking a major cost reduction in an industry whose annual sales are $4 billion. But there are cautions here, too. Vitamins and antibiotics are *now* produced via enzymes, without recombinant DNA technology, but in the same kinds of "factory cells" we've been talking about. A long, complicated series of production steps is followed within the cell. Recombinant DNA methods *could* speed up some slow steps, perhaps even radically, but tinkering with such complicated processes could result in no net benefit in some cases because of unexpected side effects. In other words, the potential for genefacture in this area is big, but the near-term uncertainty is also high.

Industrial microbiology already encompasses more than drugs. The amino acid industry is worth $1.7 billion a year, even before genefacture streamlines its methods and improves products. Amino acids, the building blocks of proteins, are valuable nutritional supplements for animal feed, but the best known of them, glutamic acid, is eaten by humans. Better known by the chemical name of its salt, monosodium glutamate, MSG is consumed as a meat tenderizer and flavor enhancer at the staggering rate of 300,000 tons per year, and it too is produced via fermentation. Aspartame, marketed by the G. D. Searle Company as NutraSweet, is produced from the amino acid phenylalanine. A true child of biotechnology, aspartame has enjoyed phenomenal sales since its introduction a few years ago, and it has spawned a gigantic headache for one gene-splicing company and an equally gigantic lawsuit in the process, as we'll see (Chapter 5).

The relationship of industrial fermentation to recombinant DNA is as close as hand in glove: the bacteria or other microroganisms redesigned through genetic engineering must be grown in huge quantities in order to mass-produce the proteins they are making, and generally that is done by a form of fermentation.

Industrial Chemicals

For years the organic chemists working in industry to develop plastics and other products have been aware of an embarrassing inefficiency in their pathways from starting ingredients, or "feedstocks," to the cooked-up end products. In the *Scientific American* issue mentioned earlier, microbiologist Eveleigh quoted from the long-standing "organic chemist's ode":

> Lord I fall upon my knees
> And pray that all my syntheses
> May no longer be inferior
> To those conducted in bacteria

The prayers are being answered—no small irony here—with the newfound cooperation of the bacteria themselves. The greatest potential for biotechnology in the long term may prove to be production of major industrial chemicals that are not the product of brute-force technology. Essentially, that means replacing energy-hungry inorganic catalysts with enzyme catalysts of incredible specificity. Scott King, one of the earliest stock market analysts to specialize in biotechnology, pointed to a huge chart on his office wall displaying the known enzyme chemical pathways. "That's what biotechnology is all about," he said. "Whatever major [activity] is going to come out will involve a better understanding of enzymes."

Enzymes act principally as biological catalysts of chemical reactions. A *catalyst* is a substance that speeds up specific chemical reactions, but remains unchanged itself at the end of the reaction; for that reason, the same catalyst can be used again and again. Effects of catalysts can be dramatic. Some chemical reactions that could take longer than a lifetime to occur without a catalyst are completed in a few seconds in the presence of the right one. Catalysts are a mainstay of chemical manufacture, but most catalysts now used are not enzymes. With the availability of cheap enzymes through genetic engineering, however, much of the chemical industry may "go enzymatic" by the end of this century.

To oversimplify the valuable specificity of enzymes, imagine that every time you mix chemical A with chemical B, you get a giant batch of C. But just by chance, a few undetectable molecules of another *very* useful chemical, D, are formed on the rare occasion that A and B join up in a slightly different form than C. Small quantities of other AB combinations will form by chance, too. Chemists work to find ways by which, sometimes at high temperature and high pressure and usually with the help of a catalyst to trigger the reaction, they can consistently make batches of D; and once formed, the molecule may be very stable. *Enzymes* are biological catalysts, and, further, they can carry out the highly *selective* manufacture of given molecules that may ordinarily be rare. If an enzyme makes D, it makes D alone, without by-product. That's what we mean by *specificity*. The synthetic chemical industry, from plastics, adhesives, dyes, and solvents to food additives, is based on the properties of catalysis: muscling molecules into different shapes than they take in nature, or producing them in much larger quantities than they are naturally found.

The association of industry with power and brawn is not misplaced. Molecular bonds of industrial feedstocks are often strong. Breaking them requires a lot of energy, which is lost as heat when the bonds are re-formed into their useful combinations. What is left behind often includes polluting residues that are difficult or impossible to get rid of.

By contrast, the enormous variety of molecules that compose the human body are basically built out of carbon, hydrogen, oxygen, nitrogen. The combination of these elements into various molecules is determined almost entirely by the catalytic action of enzymes. That's an indication of how important this class of proteins called enzymes is, and how powerful. Many of the molecular bonds of living things are also very strong.

Enzymes mediate the breakdown and reshaping of all these chemicals in ways that require much less energy. They are nature's answer to the blast furnace and the ten-ton press. That's why molecular biologists believe that the opportunity to put enzymes to their own uses will change the face of industry, as they replace energy-hungry inorganic catalytic processes with energy-efficient enzyme catalysis of incredible specificity.

Such a change will not represent the first exploration of producing chemicals via biotechnology. Pasteur's work led a German biologist named Eduard Buchner to discover cell-free metabolism. That is, he learned how to obtain the products of cellular metabolism without anything but a mix of enzymes and their "food" as feedstocks. During World War I, Germany used industrial fermentation to produce glycerol for explosives after the British blockade cut off supplies of the usual feedstock. As much as a million pounds of glycerol a month were produced by fermentation. Meanwhile, the British were short of acetone, another vital ingredient in munitions manufacture. Chemist Chaim Weizmann developed a method of producing acetone biochemically (he later became first president of Israel). Weizmann's method for producing acetone was used for many years after the war, losing out finally to processes based on petroleum—then so cheap and in seemingly inexhaustible supply.

Some genetic engineering firms believe enzymes derived from microorganisms will soon become tools in the $20-billion-a-year plastics industry. Several plastics are made by using a class of base chemical known as alkene oxides, now derived from petroleum.

Alcohol Of all the chemicals that can be made by genetically engineered organisims or via the enzymes those organisms produce, perhaps none is more immediately valuable than plain ethyl alcohol, that ancient beverage and century-old antiseptic that we have recently come to consider seriously as a fuel. We tend to ignore the fact that alcohol is already the most important industrial solvent after water and a major starting material for making industrial products. Synthetic ethyl alcohol has a market value of some $315 million a year. Naturally fermented alcohol—which is used for beverages as well as for industry—has a market of more than $600 million a year.

Ethyl alcohol, also called ethanol, can be made from waste paper, wood pulp, and other agricultural scrap. Such a conversion also has a social benefit—removing waste—that can defray costs of what might otherwise be an uneconomical process. The 1983 production *cost* of alcohol from scrap, according to a U.S. Army study from Natick, Massachusetts, was $1.38 per gallon.[3] The Cetus Corporation conducted a pilot study with National Distillers concluding in 1984, aimed at making ethanol from scrap. Beginning in that year, however, gasoline prices dropped and only recently leveled off; in the near future they are not expected to rise near the prices of the oil-embargo days, so alcohol is not yet competitive as a fuel.

But consider the production process. A mash of cellulose-containing products is treated with a variety of enzymes called cellulases, to yield glucose. Yeast is then used to convert the glucose to alcohol. The army study set the price of the cellulases, which are made by a fungus, as 57 cents of the $1.38 cost per gallon. But some scientists believe that genetically engineered bacteria could manufacture *better* cellulases *faster* than the fungus does, so that enzyme costs might be cut from 57 cents to 10 cents per gallon of alcohol. That could cause a radical change in the use of alcohol and the cost of the products in which it is a component.

Further enlivening the possibilities for commercial alcohol genefacture, biologists are trying to accomplish fermentation with a class of bacteria called thermophiles, which thrive at temperatures of around 158 degrees Fahrenheit (70 degrees Celsius). That temperature is not far below alcohol's boiling point. Much of the cost of alcohol production is in distilling it, and that means heating up the mash so the alcohol boils off. Currently, the alcohol-producing yeast die off at the high temperatures needed for distillation, so they have to be regrown for every batch of alcohol produced. Alternative: genefacture the alcohol in heat-loving (thermophilic) microbes at high temperature, so that with only slight suction the alcohol would boil off as it was being *fermented.* The thermophile would remain alive to produce more alcohol indefinitely.

Much ethyl alcohol is now derived from corn or grain, and both ethical and practical problems arise in discussions of fuel alcohol. It has been estimated, for example, that to run only 20 percent of America's cars on alcohol would require turning the country's entire corn crop in a *record* year into alcohol. Analyst King scoffed: "There will never be enough alcohol to use as a serious fuel unless we give up eating." The ethical qualm of turning a hungry world's food supply into fuel can be resolved by using agricultural waste, but the staggering scale of fuel-alcohol projects remains a problem. And there are other, technical problems as well.

For example, in the July 31, 1981 issue of *Science* magazine, four scientists analyzed the use of "biomass"—any mass of organisms or their debris—as a source of chemical feedstocks.[4] They pointed out the advantages of reduced dependence on imported oil and sharply cut atmospheric pollution, and noted that existing industries would be willing buyers; so gene-splicers would not be at a loss to sell their product. But the scientists also warned that the transfer of the new technology does not look at all simple. The current method of producing ethylene, for example, yields propylene, another valuable chemical, as a by-product. If we want to make ethylene exclusively from biomass, a way will have to be found to make propylene economically as well. They estimate that alcohol will not enter the feedstock market until its cost has been cut fivefold. Impossible? No, but uncertain. The social "redemptions" of using biomass may make some economic loss worthwhile, however, especially as part of a waste-conversion project.

Still, we must not overlook the fact that alcohol is already used as both a fuel supplement (in gasohol) and an industrial solvent. Genetic engineering may begin to make itself strongly felt as one producer even before it replaces other methods, although that probably will require a new rise in oil prices. Similarly, most of the industrial enzymes may soon have economical, revenue-producing uses, even if they do not reshape a whole industry for some time.

Glucose isomerase is the key enzyme in converting glucose—a relatively unsweet sugar found in both grapes and blood—to fructose, a very sweet sugar. It would take less of such a super-sweet sugar to bring a beverage or commercial food to the right taste level, so most genetic engineers believe pure fructose would be much in demand. For a time, "high fructose corn syrup" (HFCS) was used extensively in industry because, even though the overall mix of sugars in it only brings HFCS to the sweetness level of table sugar, it was far cheaper. In 1983, for example, Coca-Cola used HFCS for about half its sweetener in noncola drinks, and used corn syrup rather than sucrose or table sugar for all its U.S. marketed drinks. Dropping raw sugar prices have cut the market for HFCS sharply.

For diet sweeteners, aspartame (NutraSweet) has all but taken over the market. Aspartame is a chemical combination of aspartic acid, which has been made cheaply without recombinant DNA processes, and phenylalanine, which several firms are trying to make more cheaply and efficiently, in bacteria. Many people think purely of sophisticated uses for genetic engineering, but here is one major food industry application that has gone almost unnoticed.

Other industrial products The OTA study mentioned earlier predicts that genetic engineering will be used to convert sewage, landfill garbage, and agricultural waste to methane, although all methane will not necessarily be genefactured. The current value of industrial methane, however, a major feedstock that ultimately could be made by biotechnology, is $12.57 billion. The OTA forecasts that some of the hydrogen and ammonia needed for major industrial production could be produced through genetic engineering within fifteen years. Methane and alcohol production unfortunately are two areas in which energy economics are playing a more decisive role than scientific abilities.

Leading genetic engineers believe the technology to produce methane from landfills and sewage or to produce alcohol from agricultural waste could be in place within a few years. But dropping oil prices through the early 1980s are keeping these sources of fuel and industrial feedstocks noncompetitive. We will pay for that shortsightedness, as Glick noted, by not having alternate technology in place when the next oil crunch inevitably strikes.

Genetic Farming

Future changes in the way we eat, farm, and think of crops and cattle through biotechnology are staggering in scope, and they are only a few years away. In this area, many if not most of the problems and uncertainties are scientific rather than economic. But the form those changes take should be the one we've discussed since the outset: first, replacement of the way we now do things with new ways; later, new inventions; and finally perhaps even changes in context, one of which we'll suggest at the close of this chapter.

In genetic engineering's first uses, instead of merely crossbreeding crops to grow taller or increase their yields, or crossbreeding cattle to give more milk or yield more meat, we will genetically engineer them to do so. Plant and animal breeding are laborious, time-consuming, and very expensive. We may be near the time when crops can have vital human nutrients that they lack spliced into them. For example, scientists are trying to splice into corn the genes calling for high production of the amino acid l-lysine, which humans need. Since corn is a major staple in some Third World countries, improving its food value would boost nutrition without requiring increased supplies.

Modern farming is harnessed to nitrogen fertilizers, which require large amounts of petroleum products in their production. Farmers in the United States spend $1 billion a year just to fertilize corn, for example. It all seems like a mistake of evolution—fertilizer is needed only because most plants cannot "fix" nitrogen in their root systems to manufacture key protein-building amino acids, even though nitrogen makes up the bulk of the air around them.

But evolution played a lucky trick on some plants. Legumes like beans, peanuts, clover, and lupine need no chemical nitrogen source and not only grow without fertilizer, but enhance the soil for anything planted after them—a fact that has been known since Roman times. The benefactor is actually a bacterial infection; *rhizobium* bacteria infect the roots of legumes in a classic example of symbiosis. The bacteria get their nourishment from the plants, then give the plants nitrogen-rich ammonia in return. Some other plants are also able to fix some nitrogen through the intercession of other bacteria, but none so far as successfuly as legumes.

It is often noted that the entire food cycle ultimately depends on plants, which get their energy directly from the sun and use it to build sugar in the photosynthesis process. Animals lack this ability, and must therefore eat energy-giving substances, such as plants, but at the same time they derive many needed amino acids from this food. On the other hand, plants cannot get nitrogen-based amino acids they need through photosynthesis. That's why the *rhizobium* infection is so useful, providing the nitrogen foundation for the amino acids, and why a long-term major goal of "agrigenetic" engineering is to teach plants to fix their own nitrogen.

Biologists also hope to splice in genes that increase the efficiency of plants in converting sunlight to sugars, which could lead to vastly increased plant yields without an increase in fertilizer demand.

Tinkering with a plant's genetic information could produce crops that are able to grow in salty environments. That would mean food crops might be raised on currently nonarable land, such as alkaline deserts. Some plants can already live in such deserts, and scientists think that eventually they may be able to splice the genes leading to that ability into food crops. In the nearer term, plants may be engineered to produce natural pesticides, cutting food production costs and eliminating a major health hazard and cause of pollution.

Futurist Alvin Toffler[5] has pointed to the exponential rate of change that confronts us in the twentieth century—a rate of change that makes present methods of prediction obsolete. Applying a similarly rapid rate to genetic engineering, one can imagine some day creating an edible plant that would grow in salt soil without fertilizer, be extremely efficient at photosynthesis, and have other qualities that would make its waste parts ideal for conversion to alcohol. This would be a good example of "synergy," in which advances reinforce one another to accelerate even the rate of change.

A MEASURE OF OUR RATE OF CHANGE

1982. Molecular biologists are speaking with great optimism about the promise of recombinant DNA in farming, but it is a future that will not *begin* for as long as ten years. Cloning new genes into plant cells might require overcoming a host of problems, problems that either were long solved in bacterial genetics or never existed. Not one gene has ever been spliced into a plant so that its protein product was expressed.

Five years later. Several genes have been genetically engineered into plants, and their *expressed* protein products appear to confer a wide range of agriculturally useful properties.

Here are some of the achievements of this brief period.

- CalGene, a biotech startup and one of the largest devoted to plant genetic engineering, has found a way to beat a herbicide. CalGene cloned into tobacco plants an enzyme that makes them resistant to the herbicide Roundup, made by Monsanto. Significance? Roundup is so potent a plant killer that it can be used only to clear fields; it is utterly unselective. But if valuable crop plants could be engineered to resist Roundup, then a farmer could spray it in planted fields to destroy weeds that compete for precious fertilizer, water, and nutrients. CalGene's strategy takes advantage of the fact that Roundup kills plants by disabling an enzyme they must have to survive. The company cloned into the plant the gene for a similarly functioning enzyme from a bacterium that is not harmed by Roundup. Why tobacco is used for these and other experiments is explained at the end of this section.
- Using a slightly different strategy, Monsanto cloned Roundup resistance into petunia and tomato plants as well as tobacco. Monsanto engineered the plants to make thirty to forty times as much of the critical enzyme as they normally do, so they "swamped out" the effect of the herbicide.
- Agrigenetics cloned the rich storage protein phaeseolin into tobacco. Increasing the storage proteins in crop plants instead of tobacco could markedly increase their nutritional value.
- Monsanto, BioTechnica, and others have cloned into petunias and tobacco the gene conferring resistance to a toxic chemical called kanamycin; just as important, they have shown that resistance is inherited by successive generations grown from seed. This indicates that it would be feasible to produce large quantities of genetically engineered seed for farming. Kanamycin itself is not important, but the demonstration is a vital step in taking plant recombinant genetics out of the laboratory and into the field.
- Washington University and Monsanto have genetically engineered a virus protein gene into tobacco—that is, it is a gene that expresses a viral protein inside the tobacco plant cells. The surprise: the protein confers *resistance* to the virus, as if the plant had been vaccinated against it. The very latest questions in agricultural recombinant DNA concern whether such a strategy might be used to "vaccinate" a wide range of key crop plants against a wide range of viruses. We will discuss this

cutting-edge plant engineering at the end of the book, after the science of genetic engineering is clearer.

- Rohm and Haas spliced the gene for thuringiensis toxin into tobacco. This toxin is utterly harmless to humans, but it is fatal to insects, eating away their stomach linings. Conceivably, a new generation of pesticides might salvage millions of dollars in lost food crops without adding toxic waste to the environment.

Monsanto is involved in nearly all the pioneering work mentioned. In fact, this giant company stands out even among other large corporations in funding and carrying out biotechnology research, not only in plants but in bacteria and animals. Originally a producer of chemicals and chemical feedstocks, Monsanto recently bought G. D. Searle, paying $380 million to acquire its pharmaceutical markets. As we will see as this story unfolds, the Gene Age began in university laboratories, moved out into small entrepreneurial companies, and is now blossoming as large corporations such as Monsanto, Johnson & Johnson, Exxon, and Anheuser Busch pump billions of dollars into its creation.

Plant genetic engineering, only in its infancy, already has drawn more critical attention than recombinant DNA in any other area—including human and animal genetics—and that largely because the immediate goal is to place the product outdoors, in the environment. The potential hazards—real and imagined—in such a release are covered in Chapter 8. Even with potential regulatory problems overcome and field tests underway, it still will be four to eight years before genetically engineered seed is sowed across large acreages, and still a few years after that before such major crops as corn, wheat, and soybean are affected.

Why waste energy to save that pernicious weed, tobacco? For very good reasons. To a scientist, tobacco is the *E. coli* of plants. It is easy to grow a whole tobacco plant from single cells, and much harder to do in other crop plants, but scientists hope to overcome that difficulty soon. Further, for a variety of reasons, tobacco mosaic virus is the most studied of plant viruses. So tobacco and its close relatives petunia and tomato are the early favorites for Gene Age plant experiments, but they're really just launch pads for developments in major crop plants.

Engineering in animal DNA offers high technical difficulties, but it raises some serious social questions as well. Are we prepared, as one writer put it, to see a field of six-legged sheep or chickens, created to maximize leg of lamb and fried chicken production? In response to such comments, the University of Ohio's Wagner has said that Holstein cows might already qualify as such unnatural animals. They are milk machines, the result of years of painstaking crossbreeding for increased milk production. As mentioned earlier, Wagner's is one of several teams worldwide trying to develop "transgenic livestock" that might produce more efficiently as food sources and that might be engineered to yield valuable drugs—made not in bacteria but in cattle.

Tools Change Man

Before genetic engineering ever produced its first major product, it had an explosive impact on science, business, and the very university system that nurtured it. This book is about those varied impacts and their different forces, as well as about genetic engineering itself. Before a dollar was realized from genefacture, an array of companies had lined up with expectations unmatched since the days of land rushes, and a variety of corporate structures as different in type, size, and plan as could be found in any collection of Oklahoma "Sooners."

Major drug houses and chemical and energy companies have poured hundreds of millions of dollars—by some estimates $1 billion—into an industry that still has slim profits. Lone venture capitalists plow scattered millions into small firms ranging from elite partnerships of science's Nobel laureates all the way down to mere mailing addresses representing only, in the words of one entrepreneur, "a man and a boy." Other firms have tried to bridge the gap between university and business, setting up nonprofit foundations to spur scientific research, while contracting for purely practical work they see leading to profits. Several firms have gone public, with mixed ambitions and mixed results. Each has a different notion of what will bring success in the form of a balm for humanity and enormous profits for itself. Yet this very diversity boosts the chances for a successful future of the industry even as it increases likelihood that some firms will founder.

Already there have been casualties. Company stock prices have risen wildly, only to plummet for reasons difficult to assess. A major corporate undertaking, DNA Science, Inc., sponsored by the prestigious E. F. Hutton Company, fell apart before it got off the ground, again for reasons not clear.[6]

E. Russell Eggers, the president of DNA Science and the former chief executive of both Bendix International and the Loctite Corporation, said in an interview before the bottom fell out, "We're up against an agonizingly difficult transfer of technology" from laboratory to marketplace. "When you mention competition, most of the new companies talk about each other. But they are not competing with each other and won't be for a long time to come. They're competing with the old technology. It's time-tested, its pathways and steps and incremental costs are known down to the penny." That transfer of technology does not come when science or the new technologists and business people are ready, Eggers says, but when "the factors that underpin the old technology begin to change." Spoken six years ago, the words remain as true today, and as useful in explaining difficulties that always will be encountered by an industry always banking on the new.

Like the *Science* authors cited earlier, former Rhodes Scholar Eggers believes those factors *are* changing, but not as quickly as many would like. Some saw the failure of DNA Science to get off the ground as the beginning of a long-predicted "shake-out" in the industry, when the end of easy money would cripple all but a few firms with either heavyweight and firmly committed backers or enough investment money in the bank to survive. But that did not happen. As we see in Chapter 5, the genetic engineering industry fell somewhat out of favor with investors, though never to the degree financial observers expected, only to emerge in 1986 in vogue again. There have been other casualties. Southern Biomedical Laboratories, IPRI, and Armos, Inc., filed for bankruptcy protection. Other firms were bought out by corporate giants, but that, more often than not, proved a windfall for the biotech investors and provided needed cash infusion to the company acquired, even if it meant surrendering control.

Part of most firms' *modus operandi* involves tapping the best academic scientists for research, but they must cope with complaints that this commercialism is costing universities their better scientists. A range of critics from scientists to members of Congress see the boom as a threat to a form of free intellectual inquiry that has been a mainstay of Western culture since the Middle Ages. Who will prove right remains to be seen.

In forecasting the economic future of genetic engineering, at least, Genex's Glick believes the doomsayers have read the omens all wrong. Answering those who predict long time lags before cheaper production will be achieved, Glick noted, "The conventional wisdom in 1977 was that the first [human] insulin molecule would be created in the lab in 1982. It was done in 1978 [and was marketed in 1982]. At the same time, the conventional wisdom said the first interferon molecules would be cloned in 1987"—probably the year you're reading this book. "That occurred in 1979."

His point: the entire history of modern technology is one of forecasts being outstripped by performance. Electronic calculators in 1970 added, subtracted, and divided, measured one foot by one foot, and his own early model cost about $500. By the time Glick was crystal-ball gazing for the OTA, for under $200 he could buy one that carried out so many functions its ancestors would have filled a good-sized room. You can buy that one today for less than $20.

Analogously, sequencing genes—figuring out what order of base units occur in a given gene—was an undertaking of staggering proportions in the 1960s. It took five scientists as long as ten years to sequence DNA fragments 200 bases long. In the 1970s it took five people one year to sequence fragments hundreds of times that long, but today one person can figure out a longer fragment in less than a year.

Abdul Ali, a scientist with BioTechnica International, notes that the rate of sequencing genes has remained flat for some time—at about 1,500 to 2,000 bases per month for one worker, with variations depending on how complex the gene is to analyze. Now the limiting factor in understanding gene structure appears to be not the time consumed in the sequencing experiments, but the time it takes the scientist to analyze the results.

In mid-1986, Leroy Hood and his colleagues at the California Institute of Technology announced invention of the Sequenator, a laser reading device that automates the sequencing process and should take much of the drudgery out of the job. Ali noted that the automated device also does not require radioactive "labels" on DNA fragments, meaning that sequencing now should require no protective shielding and leave not even low-level radioactive waste for disposal. But to speed up the understanding of genetic structure, Ali and others believe, will require a major breakthrough, a whole new means of sequencing and interpreting.

The human genetic complement is so huge that a single gene, itself complex, can be compared to a chapter of a book, the book just one in a multimillion volume library, and that library is the entire human genome. As recently as two years ago, the suggestion that the makeup of that vast gene library would ever be known met with skepticism from many top scientists. Now, the major debate concerns whether the project should be undertaken now or delayed. Many leading scientists are calling for a federally backed project to map out the entire human blueprint, making it radically easier to pinpoint genetic defects of any kind, or to tailor drugs for particular ailments, or to understand aging, or fetal development.

"The cost would be small compared with, say, NASA's budget," one cancer researcher told us recently. "The benefits would be far more than from space travel."

But inventor Hood and the same scientists who want to undertake this project reached consensus that it should not be started yet, probably not for another five years. By then, they believe, breakthroughs will have occurred that will make the project move radically faster.

New discoveries accelerate synergistically, each boosting the others' impacts. That, at least, is one theory of how quickly tomorrow will come. But regardless of speed, surely no means of reshaping the future offers more intriguing food for thought than the reshaping of society by its own inventions, and as a hypothetical example of how this might happen in the Gene Age, we offer a case posited by the *Science* authors referred to earlier.

Genetic engineering may produce industrial feedstocks with great energy savings, but that is only phase one in the progressive change tools bring to their inventors. Of at least equal importance farther down the road is the fact that these new feedstocks might be *grown*. Agriculture is not only labor-intensive, it is unskilled-labor-intensive. The agriculture needed to feed manufacturing plants would not have to be either vast in scale or centralized, but could be dotted over wide areas in many countries, as could the manufacturing plants for the feedstocks. It will be ironic—but a happy irony—if this most sophisticated of all sciences could lead to a system that would bring major rewards to unskilled farmers in underdeveloped countries by allowing them to remain farmers, but to reap some of the rewards of industrial society by growing the raw materials of industry. This kind of change would mark the third phase of a bio-revolution, the change in the shape of society and the relationships of its members.

NOTES

[1]*Impacts of Applied Genetics,* referred to thoughout Chapter 1, was published by Congress's Office of Technology Assessment in 1981 and is for sale by the Superintendent of Documents, U.S. Government Printing Office, Washington, DC 20402.

[2]*Scientific American,* special issue on "Industrial Microbiology," September 1981. Reprinted with permission. This issue contains perhaps the most comprehensive look at industrial uses of genetic engineering so far.

[3]"Cellulases: Biosynthesis and Applications," by Dewey D. Y. Ryu and Mary Mandels, in *Enzyme and Microbial Technology,* Vol. II, 1980, pp. 91–102, contains Natick's alcohol production estimates. Reprinted by permission of the publishers, Butter & Co. (Publishers) Ltd.

[4]"Biomass as a Source of Chemical Feedstocks: An Economic Evaluation," by B. O. Palsson, S. Fathi-Afshar, D. F. Rudd, and E. N. Lightfood, *Science,* 31 July 1981, pp. 513–517. Copyright 1981 by the American Association for the Advancement of Science. Reprinted with permission.

[5]Alvin Toffler, *Future Shock* (New York: Random House, 1970).

[6]"DNA Science Inc.: First Casualty in the Biotechnology Derby," by Colin Norman, *Science,* 4 September 1981, pp. 1087, 1088, 1090. Copyright 1981 by the American Association for the Advancement of Science. Reprinted with permission.

2 GENETIC ENGINEERING: THE SCIENCE I

WHEN scientists say that DNA is the secret of life, they mean that the awesomely long yet simply built molecule contains the blueprint for each particular living thing in which it is found, a blueprint not only for creating cell parts out of the building blocks we obtain from food, but for assembling parts into the whole and for maintaining the organism in its living state.

The surprisingly simple relation of DNA's form to its function, which we will discuss shortly, has awed scientists since those who discovered its shape first marveled at it thirty years ago. In science as in art, the truly powerful concepts are those which emerge with clean-lined simplicity from dissonance and complexity—and there's plenty of complexity to be smoothed out in learning how molecular biologists have harnessed the power of DNA to turn billions of cells into factories for chemicals and drugs.

In function, DNA is similar to the master blueprint for a planned-community of houses. But it's not a drawing. This blueprint is a set of instructions communicated in a unique code whose "alphabet" uses only four submicroscropic units as letters. Billions of these molecular code-letters could be strung out end to end on a postage stamp.

In this chapter, we look at how the instructions contained in DNA are put into action in the living cell, in particular, in the organism *Escherichia coli—E. coli* for short. This simple, one-celled bacterium has long been the favorite subject for microbiological study. We concentrate on those aspects of an *E. coli* cell responsible for reading DNA langauge and then carrying out the instructions telling the cell how to grow, divide (that is, produce a living copy), and maintain itself in a healthy, living state.

Virtually everything we talk about here applies to the human body and its cells as directly as to bacteria or any other living organism. Our own DNA replicates in much the same way as the DNA of *E. coli,* and with the assistance of enzymes our DNA orders the construction of our own body-building proteins using the very same code-language. Among the many wonders of molecular biology is the fact that the genetic code is universal. No Tower of Babel here: the same "words" that instruct the *E. coli* to add another particular amino acid to a protein chain would order up the same amino acid in a honeybee or a man.

For the last vital cellular requirement mentioned above, maintaining itself in a healthy living state, a cell must manufacture a myriad of chemical products, among them many different kinds of proteins, including enzymes, structural proteins, and sometimes hormones; polysaccharides (complex sugar-type substances that are the "batteries" of living cells); and other important metabolites, including such organic chemicals mentioned in Chapter 1 as vitamins, amino acids, and alcohols. Basically, all organisms, whether microbes or humans, are built out of proteins. Enzymes are a type of protein that carry out maintenance and "worker" functions in the body, and hormones are proteins that carry out more complex regulatory functions.

The task of genetic engineers is to subvert this natural chemical manufacturing process to their own ends in order to turn the cell into a factory that makes commercially useful products. Of major importance are the problems that crop up when the genetic engineer takes over *E. coli*'s processes, because these problems determine how long it will take to commercialize potentially useful products. We'll also consider organisms other than *E. coli* that might have important uses as hosts.

Everything in this first science section is vital to understanding genetic engineering, and everything in the process of DNA function and cellular life that does not relate has been eliminated. The later science sections range a little wider over the field, discussing some of the specific tasks of genetic engineers and some of their latest tools. Finally, the last section (Chapter 9) continues the discussion all the way through manufacture of a protein and explains some of the pitfalls in commercial "gene-splicing." The last science section also includes a short course in the genetic language far simpler than French or German, but one in which the most profound secrets of life are written.

DNA Through the Schematic Looking Glass

DNA is a chain molecule containing millions of "links," but even such a long molecule is too tiny to be made visible except with the most modern techniques and the most powerful electron microscopes (see facing photograph). To offer some idea of the incredible proportions involved, an average bacterial DNA molecule is about one twenty-fifth of an inch long, and that certainly is *long* enough to be visible; but its *width* is only one millionth of that. By comparison, if you imagine that molecule as two intertwined chains whose links are an inch wide, the chains would be more than fifteen miles long. The DNA found in every human cell would be nearly two thousand times that long—nearly long enough to circle the globe.

The puzzle of the structure of DNA was finally cracked by James Watson, an American, and Francis Crick, a Briton, at Cambridge University. That discovery is considered the major milestone in modern molecular biology, and it earned Watson and Crick the Nobel prize in 1962.

The molecule resembles a ladder that has been twisted to resemble a spiral staircase. To visualize this all-important structure, look at it under the "schematic microscope" we've created in Figure 1, on page 39. The schematic microscope will allow us to see biological life at various levels of magnification in artist's renditions. Each of the DNA representations shown is used by scientists to highlight different sets of qualities. At the left of Figure 1a is the double helix, closest to DNA's natural configuration. At the right, a major activity is shown: the unwinding or unzipping of the two strands, the first step in DNA's replication. Of all DNA's characteristics, none is more amazing than this ability of a lifeless chemical molecule to reproduce itself, as we'll witness in a moment, and that ability is directly related to its structure. The unwinding of the two DNA strands can occur because the two "backbones" you can see as wavy lines on the double helix are very strong; all of the identical backbone elements are held together by strong molecular bonds, just as our vertebrae are strongly joined yet

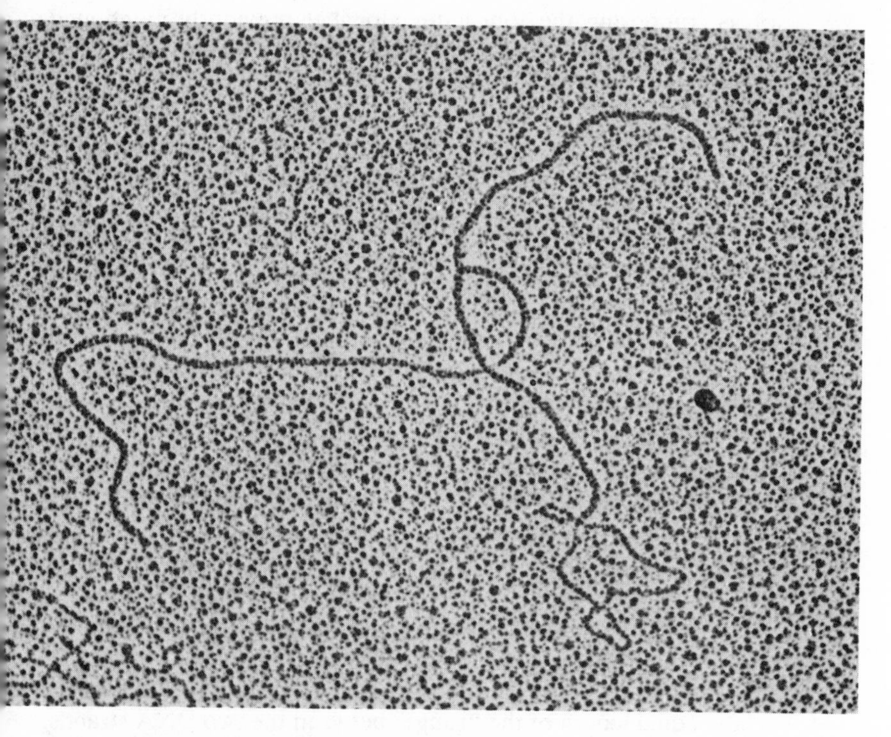

A bacterial DNA molecule magnified 80,000 times.

supple. The single steps in this spiral staircase are actually formed from two separate pieces, joined edge to edge. Each separate piece is called a *base* (solid vertical lines in the diagram). These base pairs are *weakly* joined, and when the DNA unzips, the steps split up the middle while the backbone remains intact. Figure 1a shows the strong bonds as solid lines, the weak bonds as dotted lines.

Figure 1b shows the zipped-up molecule in a different schematic, appropriately called the ladder representation, in which the helix has had the twists taken out, and Figure 1c shows the ladder at higher magnification to reveal some of the substructure. Here you can see the most important aspect of DNA from a genetic or hereditary point of view. The letters represent the bases, which are chemical compounds: adenine, thymine, guanine, and cytosine. *The order of these bases along the DNA backbone carries all the information of heredity*. The bases A, T, G, and C can be thought of as composing the four-letter alphabet previously mentioned. The order of the bases pictured in the lower strand of Figure 1c—CGAATGTTCA—then can be seen as a written instruction in DNA language that will eventually result in the ordering of amino acids into various proteins. If you were to change the order of the bases, you would change the instruction. In a living cell, DNA language is read not by eyes but by other molecules, which then carry out its instructions. Later, we'll learn how the cell reads and interprets DNA language. To get some idea of how significant these commands are, consider that the substitution of just one "letter" for another in a single command out of millions in the human DNA chain is the cause of several genetic diseases, including sickle-cell anemia.

At this magnification the extremely important principle of complementarity can be seen. The G base (guanine) on one backbone *always* bonds to a C base (cytosine) on the other backbone to form one base pair, and an A base (adenine) always bonds to a T base (thymine) to form the second type of base pair. One of Watson and Crick's major realizations leading to the unraveling of the DNA puzzle was that although the shapes of the four bases are very different from one another, the GC base pair is almost perfectly congruent with the AT pair. This is shown in the schematic by the equal length of the "rungs" between the two DNA strands.

Looking at the magnified strand, complementarity tells us that whenever there is a T base on one strand, there will be an A base opposite. That means that the two strands are not identical, nor are they simply in reverse order from one another. Each is a complement of the other: if one strand reads . . . CGAAT . . . , then the complement will read . . . GCTTA . . . as the schematic shows. Further, knowing one strand's base sequence, we can always predict the other's. The chromosomal DNA of the bacterium *E. coli* is several million bases, or letters, long, so it encodes plenty of information on how to make the cell and carry out all of its life processes.

A Molecule That Reproduces

Chemical molecules form when enough atoms of the right kind are present in the right quantities and conditions, but we know of

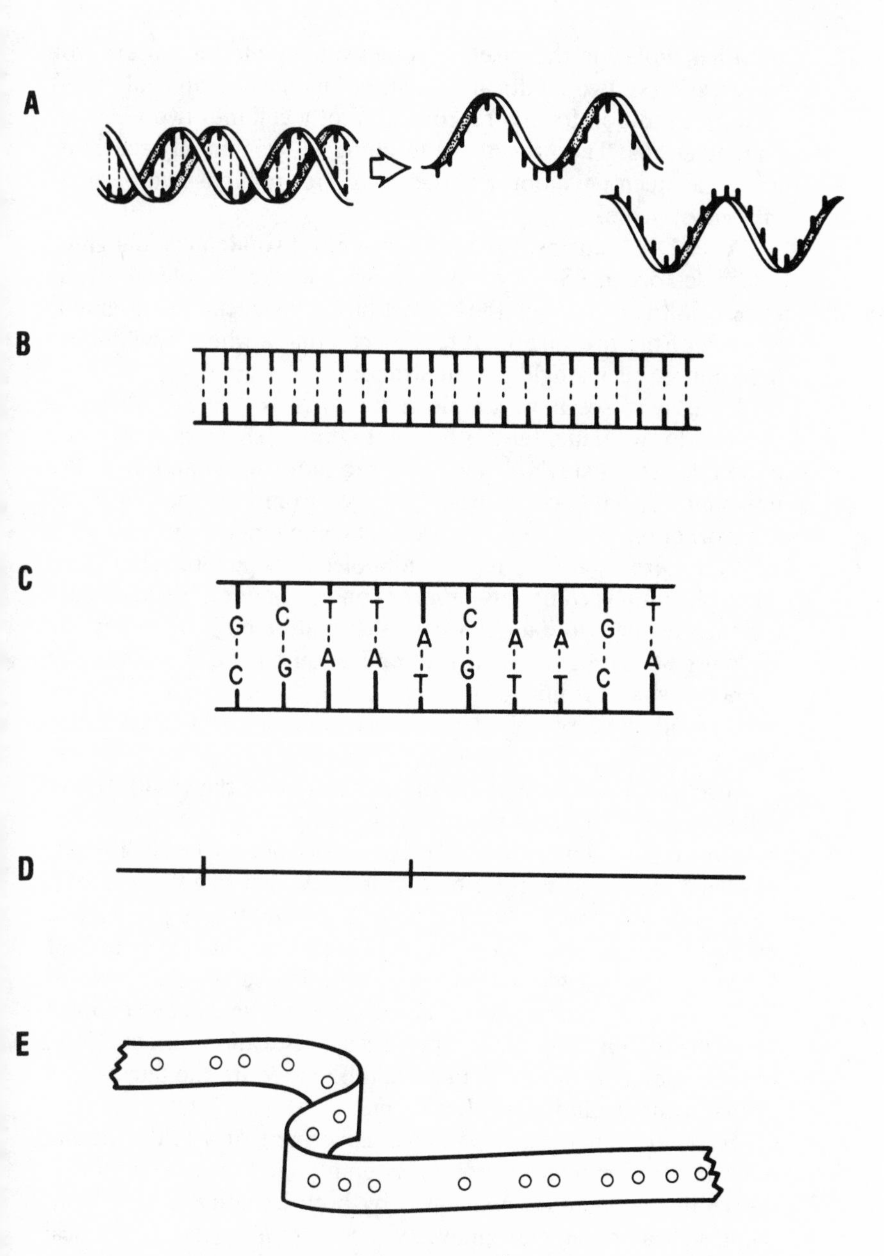

Figure 1. Five schematic representations of DNA.

no other molecule that itself becomes a template, or pattern, for the creation of two identical "daughter" molecules, an analog and vital precondition for the reproduction of a cell into two identical daughter cells. The process we're going to oversimplify schematically has been developing and evolving for millions and perhaps billions of years.

After DNA unzips, the replication into two identical daughter molecules occurs. Simply put for now, "worker" molecules (enzymes) in the cell attach the correct base-plus-backbone fragment to each of the *two* unzipped halves, creating a whole new double helix out of each single parent strand.

It's easy to see how complementarity is used in DNA's replication into two daughter molecules. Figure 2a shows a small piece of double-stranded DNA, and Figure 2b shows the strands partially unwound. From a pool of bases (derived from food the cells eat), the worker enzymes attach the correct complements shown in Figure 2c, where a partially replicated molecule is pictured. In Figure 2d, replication is complete—there is one daughter DNA molecule for each of the two daughter cells that will be built. A molecular biologist will recognize that the replication has been so radically simplified that it is not scientifically correct, but it does show how the key principle of complementarity is involved in DNA replication.

If so much is known about the duplication of the vital DNA in cell replication, it might seem an easy jump to see how the rest of a living cell is duplicated, but that turns out to be a difficult, unsolved puzzle. Molecular biologists know that to a large extent, many cell components self-assemble. This remarkable property of many biological structures can be likened to shaking a box of Lincoln Logs and tossing out completed log cabins. We have to fall back on our old answer: that self-assembly must have evolved through millions of years to its present sophistication, just as DNA, enzymes, and all the molecules we talk about are the latest steps in that same evolutionary development.

Molecular biologists also are discovering that a cell's internal environment influences protein assembly. Just as a living organism's development is controlled by both heredity and environment, so is a protein's assembly. Two trees from genetically identical seeds might look unrelated because of differences in each one's surrounding temperature, or available sunlight, water, and nu-

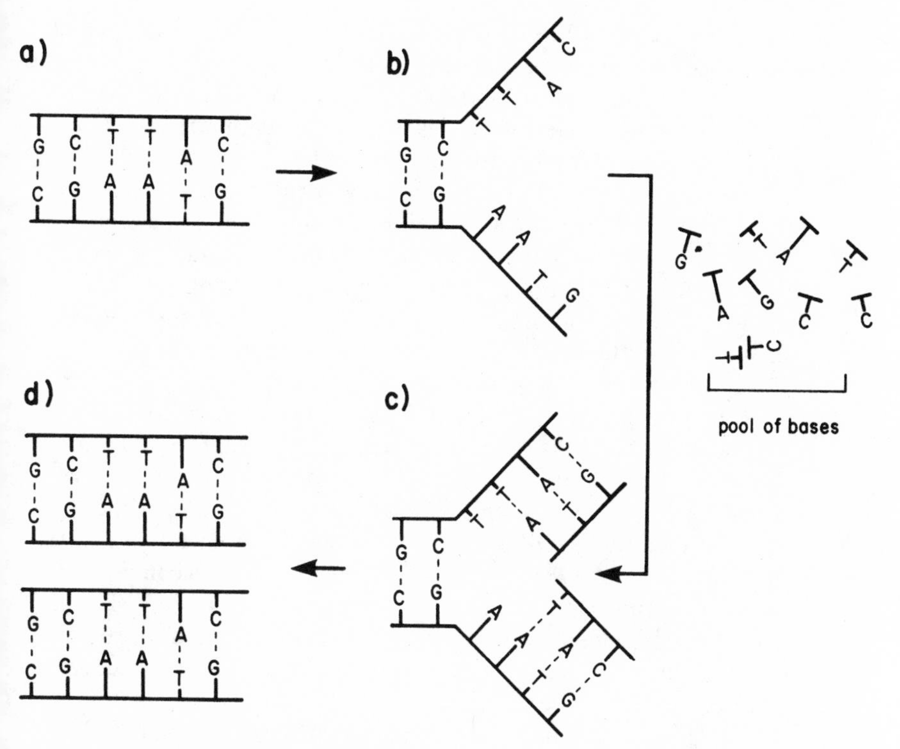

Figure 2. The replication of DNA.

trients. Now scientists have found that some animal proteins fold up into a given, highly functional shape in their "native" environments, but they fold up differently and become ineffective when grown in *E. coli.* The cause appears to be the difference in chemical environment within the two kinds of cell, leading to different chances for certain "finishing" chemical reactions to occur, for example.

How *E. coli* Makes Its Living

E. coli has several qualities that long ago recommended it to biologists for study. Although a dweller in the human intestine, it is harmless to us; it reproduces rapidly, and it can be cultured in the lab on a diet of sugar water and a few other simple, inexpensive nutrients. These virtues led to its becoming the best-known microorganism in molecular biology, and for this reason it was the first of the genetic engineer's cellular factories.

A cell's single most vital job is to reproduce, and at that *E. coli* is a champ. One cell divides into two daughter cells identical to the original every twenty minutes, under optimum conditions. That means we get four cells after forty minutes, eight cells in one hour, sixty-four in two hours and, in a mere ten hours, a teeming colony of over one billion, each identical to the original parent (in the absence of mutation). The great power of genetic engineering is tied to this wholly natural reproductive ability; if just one cell is built right, the production boom almost takes care of itself.

E. coli is unicellular, each cell living alone in its environment with no attachment to others of its type. Humans, of course, are multicellular: billions of cells of such types as skin, muscle, and blood are organized into a community, forming a (usually) functional whole. "Microorganism" means too small to see without a microscope; one *E. coli* is about 1/25,000th of an inch long and a fourth as wide. An *E. coli* cell and some of its internal components are shown under the "schematic microscope" of Figure 3. To see how the components interact to build, maintain, and reproduce the cell, consider this fable of one developer's awesome success in creating a planned community.

Let's call our protagonist Frank Lloyd Levitt, a home-builder with a terrific scheme for building whole subdivisions of identical houses with which he dreams of covering the suburban New York countryside. The key to Levitt's method is his "pilot" house; within it is the blueprint for the whole subdivision, explaining step by step what materials are needed and how and when they are to be joined. First, supplies such as block, lumber, and nails have to be delivered to the construction site.

He also needs a source of energy for construction. And once a house is built, he'll need whole new categories of equipment—brooms, cleansers, paint—to keep the place livable. An *E. coli* cell would similarly need to bring in food as a source of both raw materials and energy for construction, and later would need enzymes to "keep house."

The pilot blueprint, of course, is the most important single item, and Levitt doesn't want to carry it around and risk damaging it, so he photocopies various sections of the blueprint as needed. DNA is the cell's pilot blueprint, and the cell also has a copying

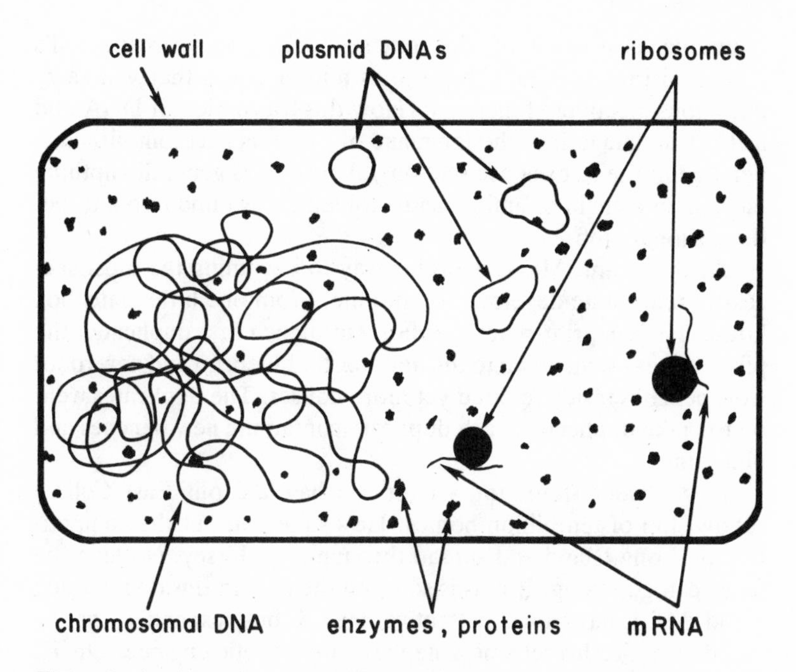

Figure 3. A schematic representation of an E. coli *cell showing the major components involved in protein manufacture.*

machine, called RNA polymerase. The "copy" of a given DNA segment this copier makes is called messenger RNA (mRNA); it's pictured in Figure 3 as a shorter curvy line than the "template" DNA.

But there is more to builder Levitt's design than the blueprint; there is also an automated workbench. Mr. L simply feeds a copy of one of these blueprint sections into the workbench; then electronic feelers scan the segment and robot arms grab the wood, bricks, and mortar. Whatever is called for in the on-site blueprint segment (mRNA) is assembled on the workbench automatically.

E. coli's robot workbench is composed of large, nearly spherical structures called ribosomes (pictured), and smaller molecules not shown, including enzymes. The ribosomes grab onto the mRNA blueprint copy. Through subtle molecular interactions, they read and execute the instructions on the mRNA blueprint at this cellular workbench. They hook protein building blocks called amino acids into chains of different sequences (much as each line of this page is unique as a chain of different sequences) to form the cell's structural proteins and the proteins and enzymes that will carry out "housekeeping" functions. (How this translation of DNA and mRNA language into the language of proteins is accomplished is detailed in the last science section.) A protein is generally nothing more than a chain of amino acids, folded and wound into a three-dimensional glob.

At this point, Mr. L is on his way to becoming the most successful real estate developer of all time. From one little matchbox house, his blueprint-reading robot workbench first duplicated the pilot house—right down to the new master blueprint and new robot workbenches inside to build yet more houses. The duplicates went on to make duplicates, each duplicate containing new benches and blueprints.

Now, twenty-four hours later, he has laid out East Coli, a subdivision of ten billion homes. He's squeezed out the competition on Long Island and spread through New Jersey. Nothing, he feels, can go wrong. But as is so often the case in this dog-eat-dog world, Mr. L has grown inattentive. He's so busy expanding that he failed to notice his scheming nephew, the genetic engineer Dr. K.

Invited to watch the amazing self-replicating subdivision, Dr. K notices that the automated workbench will build anything it reads on the blueprint. A bountiful supply of raw materials is on hand, and Dr. K realizes that if he stitches his own little blueprint segment into the master, then every house built will have an expensive sports car in the living room. All Dr. K has to do is to break out a wall, drive the car off, and he'll be in the car business without lifting a finger. By the time Mr. L finds out, Dr. K will have a fortune in once-rare cars. Among the problems: if the robot workbench uses up too much material and energy building a sports car, then it won't be able to build more houses, and that in turn will limit the number of sports cars that can be built.

In a nutshell, that is the task and one of the major problems facing the genetic engineer. *E. coli*'s ribosomes (cellular particles composed of RNA and protein that function as the site of protein synthesis) will build exactly what is put in front of them, following instructions coded on the mRNA, regardless of the origin of that mRNA. Genetic engineers have to supply the ribosomes with a good supply of the mRNA for their desired proteins, no easy task. They must make equally sure that in addition to making a lot of the particular mRNA for the desired proteins, the cell continues to make enough of its own proteins' mRNA for it to survive and reproduce.

Today's "flashy sports car" of genetic engineering is interferon, and large quantities of various types have already been produced in *E. coli.* The bacteria have no use for these animal proteins, and although we do, our cells make them in extraordinarily tiny quantities. Still, interferons are made on human-cell ribosomes just as they can be made on *E. coli*'s.

As we mentioned earlier, a gene is nothing more than a region of DNA. *Generally we think of the gene as that region which carries instructions for a single protein.* Thus, one gene codes for the production of one kind of interferon protein. Once the genetic engineer has slipped the interferon gene into *E. coli,* if he or she inserts special signals telling the cell to make many mRNA copies of that gene, the end result is a high rate of interferon production. We'll see that these signals exist and are very important to genetic engineering.

Now we're really in business: we've engineered each cell to mass-produce interferons, and hitched that engineering feat to the cell's native ability to mass-produce itself. That's why genetic engineering offers such potential for scaling up production of rare and now-costly drugs and chemicals.

Another major tool of the trade is plasmid DNA, seen as circular DNA in the schematic microscope (Fig. 3). For its function, let's head back to the subdivision of East Coli. Despite the greed of his nephew, Mr. L's subdivisions are doing fabulously in the East, and he decides to hit the Sun Belt, someplace like Arizona. But he doesn't want to build exactly the same house in Phoenix that he's building on Long Island (though many have tried). The basic blueprint is still fine, but it must be modified to take local conditions into account. Central air conditioning will be a must, for example. What he needs is something like his nephew's intrusive commands: supplemental blueprint units that will vary from one locale to another, this one adapting the houses to dry heat, that one to extreme dampness, still another to months-long cold and deep snow.

These supplemental blueprints we've envisioned are surprisingly similar to the little circles of plasmid DNA contained in many bacteria. In contrast, chromosomal DNA—the master blueprint—

contains all the information necessary to survival under "normal" circumstances. But *E. coli,* an extraordinarily well-traveled organism, might find itself in an inhospitable locale—for example, an environment containing penicillin or another antibiotic. A normal *E. coli* would die in the presence of penicillin, while one with a plasmid containing instructions for defense would survive.

Where does the plasmid come from? In a population of *E. coli* cells only a very few might contain the plasmid for resistance to a particular antibiotic. In function, the plasmid DNA would instruct the cell to make a protein that would combat the antibiotic. If you now introduce the antibiotic into the colony, all the *E. coli* without the plasmid will die. That means the future colony will be composed only of cells containing the plasmid, because they are descended from those survivors that had it. But surprisingly, cells can also pass plasmids to other cells that don't have them via a very complex mechanism. That means that when an antibiotic is introduced into a colony, some cells without plasmids will get them and themselves survive, in addition to passing the plasmid on to their descendants. Plasmids are duplicated as their hosts divide, and that is of vital importance to their use in the gene-splicing tool kit.

The genetic engineer rebuilds plasmids outside the cell, inserting into them the gene to be "expressed" (that is, made into its corresponding protein product). The engineer then places the plasmid in a broth containing "host" bacterial cells, and the hosts incorporate the plasmids into their own cells and pass the information from generation to generation as if picked up naturally.

Everything discussed so far can be summarized in two simple diagrams and as many sentences. The diagrams are standard scientific shorthand for chemical reactions and other types of transformations, showing the agent bringing the reaction about above the arrows, and demonstrating the general flow of instructions for a living cell and for our homebuilder.

DNA —(RNA polymerase)→ mRNA —(ribosomes)→ protein

Master blueprint —(copy machine)→ copy of blueprint section —(robot workbench)→ house

The genetic engineer wants to speed up any or all of those steps, and the snowballing effect of increased rates on one another is what is called "overproduction" or "maximization of gene expression." But before getting too far afield on methods, we need to look more closely at the subcellular pieces that play key roles in production or are desired proteins themselves, and to start with, we must take a closer look at DNA itself.

Cells: The Inside Story

PROTEINS, SECONDARY METABOLITES, AND OTHER CELLULAR GOODS

Some molecules such as oxygen, table salt, and water are small clusters of the same or different atoms. But most of the molecules of interest in biology are long chains of subunits, and even the subunits themselves may be fairly complex clusters of atoms. Such long-chain molecules are called polymers, and all living things contain two important classes of them: nucleic acids (DNA and RNA), the instructions for creating and maintaining life; and proteins (also called polypeptides), which we can think of as the "stuff" of life. Let's start with the nucleic acids.

As in Figure 1, each of DNA's two chains is composed of the bases A, G, T, and C. Therefore we can represent a section of chain simply as letters and dashes as in this hypothetical sequence:

. . . . . . G-G-T-C-A-A-T-G-C-T . . . . . .

In this way, DNA resembles a four-colored necklace, with each base a different-colored bead.

Proteins, too, are necklacelike chains, but their "beads" are combinations of twenty different chemicals called amino acids. The names of all the amino acids and their abbreviations are given in Table 1. Thus, a piece of a protein molecule made of a chain of the amino acids methionine, glycine, alanine, and valine can be written schematically this way:

. . . . . . Met-Gly-Ala-Val . . . . . .

The sequence in which the twenty amino acids are strung together determines the protein, just as the sequence in which the four nucleic acid bases are hooked together determines the in-

Table 1. The Amino Acids

The 20 Amino Acids	Abbreviations
Alanine	Ala
Arginine	Arg
Asparagine	Asn
Aspartic Acid	Asp
Cysteine	Cys
Glutamic Acid	Glu
Glutamine	Gln
Glycine	Gly
Histidine	His
Isoleucine	Ile
Leucine	Leu
Lysine	Lys
Methionine	Met
Phenylalanine	Phe
Proline	Pro
Serine	Ser
Threonine	Thr
Tryptophan	Trp
Tyrosine	Tyr
Valine	Val

structions of the DNA blueprint. The analogy to written language is obvious. DNA language consists of a four-letter alphabet. Similarly, protein is a language written in a different, twenty-letter alphabet. The analogy is not just a learning device. So close are the parallels with language and its uses that molecular biologists describe the flow of genetic information in a cell this way:

$$\text{DNA} \xrightarrow{\text{(transcription)}} \text{mRNA} \xrightarrow{\text{(translation)}} \text{protein}$$

The making of an mRNA copy from DNA is called transcription because it is virtually the same as transcribing notes or comments from one medium to another without altering their form.

Making a protein from the instructions contained in mRNA is called translation because it represents taking instructions written in one language (A-G-T-C-etc. . . .) and translating them into a useful form in another language (e.g., Met-Gly-Ala-Val).

We talk about the details of transcription and translation later, but these processes carried out inside the cell are the same except for details, whether the cell be *E. coli*'s or your own. Figure 3A summarizes the process in *E. coli.*

Although mRNA is a good copy of DNA, there is one difference that occurs in transcription. In place of the base thymine (T) in DNA, molecules of RNA contain a similar base called uracil (U). The difference is more in the name than the shape, for although an mRNA sequence might read -U-U-U-, its informational content is the same as DNA's -T-T-T-. Why nature chose to use this similar but different "letter" only in RNA is not fully known.

Proteins in turn can generally be considered in two classes: structural and housekeeping. "Structure" is as simple to understand as bricks, wood, or mortar, but the housekeeping functions of a protein are more complex and involve the actions vital to living: digesting food, breaking it down to cellular building blocks and/or deriving energy from it, clearing out waste, and so on. While structural proteins may have some commercial value, the housekeeping proteins are far more interesting in that respect; they are the enzymes that carry out all the cell's chemical reactions, manufacturing a tree's wood, the vitamins a human needs, or the antibiotics produced by some bacteria. These are not one-step operations, however. The enzymes or vitamins or antibiotics, produced by bacteria, operate in often-long series of chemical reactions involving perhaps dozens of different enzymes, each doing a single reaction job in its turn. Biochemists refer to these chemical reaction sequences in the cell as "metabolic pathways."

In Chapter 1 we mentioned that enzymes are good catalysts because they are so specific. Most chemical reactions involve unwanted byproducts, but one enzyme will carry out only one chemical reaction to yield a single product. That means we don't have to separate the desired product from the other residues produced along the way; we can eliminate the usually costly steps of purification. We mentioned previously that industrial force, through blast-furnace temperature and high pressure—major contributors

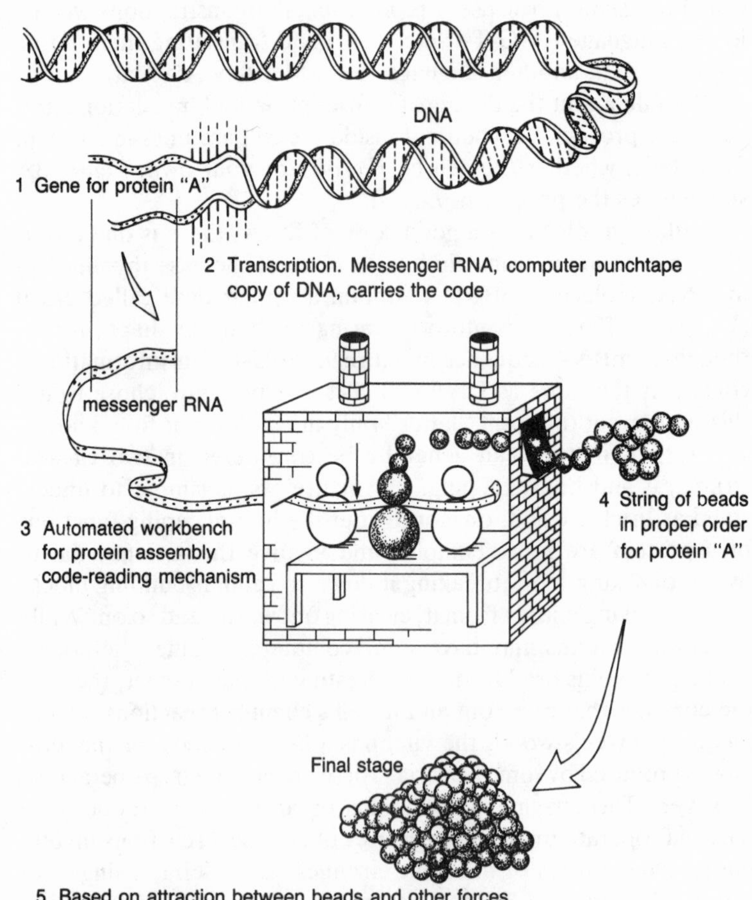

Figure 3A. Bacterium's DNA complement: "A computer code" for protein manufacture.

to environmental and energy problems—can be replaced by enzymatic power. Enzymes have to work under mild conditions because they evolved working inside living organisms.

Wood or scrap conversion to alcohol depends on the fact that some fungi and other organisms use wood as food. Most organisms can't break down the cellulose-rich wood into glucose because they lack the enzymes, called cellulases—a shame because, strong as it is, cellulose is nothing more than a long chain (polymer) of glucose molecules. If *E. coli* could be genetically engineered to make large quantities of these wood-digesting cellulases, large quantities of glucose could be made from the waste wood, scrap paper, and such farm waste as corn husks. The glucose could then serve as a food source for yeast—in yet another conversion chain—as the yeast turns glucose into alcohol as fuel or feedstock.

An enzyme-catalyzed reaction is taking place under the microscope in Figure 4. Figure 4a shows a simple chemical reaction, a *degradative* one in which a complex chemical A-B is to be broken down into parts A and B. Enzymes actually carry out both these degradative and *synthetic* reactions which, as the name implies, involve building more-complex chemical structures out of simpler ones.

In this degradative reaction, the catalytic enzyme fits its shape to the A-B molecule complex and exerts attractive forces on it, breaking the bond between the A and B parts, thus releasing the two separate molecules. The enzyme is then free to repeat the process on another A-B complex.

Figure 4b shows the catalyzing enzyme. Its most noticeable feature is its surface, which contains two pockets corresponding in shape to the A and B groups—such "lock and key" surface complements are common in enzymes. The two chemicals catalyzed by the enzyme are known as substrates. The enzyme's specificity is ensured both by the pocket shape and by varying, often-subtle attractions between chemical groups. Real molecules have incredibly complex and varied shapes, and enzymes have evolved to mirror those shapes. The actual specificity of enzymes is far more complex than we've depicted.

Figure 4c shows the enzyme bound to the substrate. Usually the attractive interactions between enzyme and substrate are weak compared to the strong chemical bonds that hold the substrate together, and that's where things get interesting. How does an

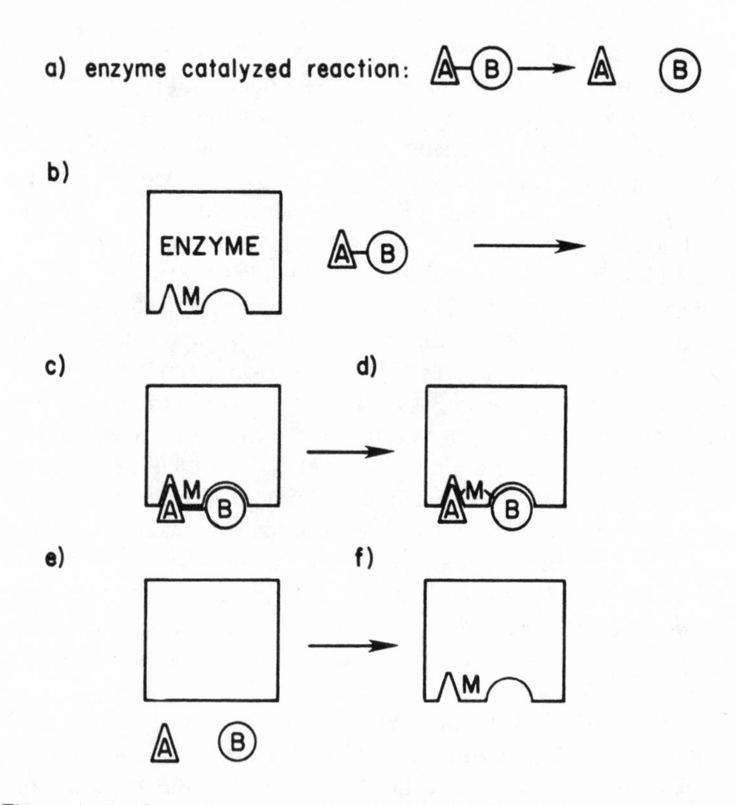

Figure 4. A schematic representation of enzyme function.

enzyme molecule break this strong bond? Think of the chemical bond as the force holding two magnets together, for chemical bonds are attractive forces, as are magnetic forces, though certainly not the same. You might unstick the magnets by yanking them apart. But an alternative would be to bring in an electromagnet; if even momentarily the electromagnet's attraction for each magnet is stronger than their pull on each other, the bond is broken. Look at the enzyme area denoted "M." If the attractive forces in this region alone are strong enough, they can help break the strong bond between A and B, as shown in Figure 4d of the figure.

But now the enzyme must release the broken pieces of A and B into the medium, and this can occur in several ways. Since the *overall* bond between substrates and enzyme is weak, the A and B molecules might just jiggle off through natural vibrations. If the overall bond is too strong for that, the enzyme might undergo a slight change of shape so its "lock" pockets no longer fit the "key" substrates. Figure 4e shows an enzyme "spitting out" A and B by flattening out its pockets, only to revert to its most stable form in Figure 4f—ready to go again.

Enzymology is among the most complex fields of biochemistry, and chaining enzymes to our own industrial production is not a simple task. We explore some interesting properties of enzymes and some of the complexities of metabolic pathways in a later science section, including some revolutionary new kinds of enzymes that promise even greater acceleration in the development of genetic engineering.

3 GENES AND GENIES

TOM ROBERTS, bent over a small plastic vial, releases a thin rubber tube from his lips and a stream of clear liquid flows from the attached pipette into the test tube. The quiet laboratory room is not new or particularly modern-looking—black stone bench tops, chart on the wall, large jars with small plastic test tubes with a refrigerator to store them. Another man works nearby, equally silent. A summer storm is blowing up outside the fifth floor corridor in a complex forming the Dana Farber Cancer Institute in downtown Boston.

Working from cookbooklike formulations in his head and out of scientific papers, many of which he helped write, the soft-spoken, self-effacing Roberts is performing an engineering feat of dazzling complexity and precision. Which one? With these same motions, Roberts is doing radically different experiments, each vital to the story of genetic engineering:

1981. He is altering the genetic makeup of a colony of bacteria so that as each cell carries out the building and maintenance jobs dictated by its genetic makeup, it will also make proteins alien to its heredity—proteins valuable to us. All the noise and hype of Wall Street are focused on quiet rooms like this since Genentech's public stock offering of October 1980 launched the pure science of genetic engineering into the marketplace.

Period piece: a Wall Street stock analyst who prides herself on critically appraising the new recombinant DNA firms argues with a scientist over whether the gene-splicing techniques are ready for the marketplace. This is perhaps the most critical question in the efforts of business and science to understand one another. "I've heard people boast that their scientists had cloned a protein," the analyst says—meaning that actual production of the desired end product had been achieved. "But when you talk to the scientists themselves, they're very straightforward. They tell you, 'Clone it? No, we've just inserted the gene for it.' "

From a practical *or* scientific viewpoint, snipping and inserting new genes into bacteria is relatively simple compared to getting the host bacteria to "read" the new gene correctly and carry out its instructions. But if you can't get the protein read out, there's no product to sell.

The scientist rejoins: "Listen, Tom Roberts can get *E. coli* to spit between 0.5 percent and 5 percent of its proteins as the desired protein. He's one of the top gene-splicers. And he can do that almost routinely: get 'expression' of the gene as protein, and also get high yields. He's just one of the best there is."

To understand how quickly recombinant DNA has moved in half a dozen years, consider that soon after that conversation, Roberts raised the percentage from 5 to nearly 15 percent. Within months, it was routine for scientists in many better laboratories to be able to get bacteria to yield that percentage. Now yields are at 20 percent and above, stretching all the way to a reported 70 percent for one protein. And as for claiming a breakthrough when an inserted gene has not even yielded its protein, by 1986 high school biology students in several areas of the country had to do better than that.

Roberts, a pioneer in these "overproduction" techniques, has not been idle.

1984. He is studying the sometimes drastic errors that take place as normal cells become cancerous—errors of the genes, the cells' inheritance, that lead them to grow out of control. Uncontrolled growth is a major feature of both malignant and benign tumors. In cancer research, the early 1980s were landmark years in discovering the actions of oncogenes—cancer genes—with much of the leading work done across the Charles River in Cambridge, in the M.I.T. laboratory of Robert Weinberg. Without the tools of genetic engineering, none of these discoveries could have been made (see Chapter 10). Bent over his stone bench in 1984, Roberts is building an experiment to learn how one particular oncogene called *src* is turned down its deadly pathway.

1986. Basic research has taken great strides as laboratory scientists like Roberts have learned to better use an enzyme called reverse transcriptase. As the name implies, this molecule allows molecular biologists to reverse the transcription process. Now when they see the results of mutation in a gene, they can trace their way back up the stream of gene commands to the mutated gene itself, to find out what changes in it led to the drastic or beneficial results.

Here in his "shop," Roberts carries on a variety of projects that seems too broad for one engineer. In addition to studying the behavior of cancer genes, he recently cobbled together key genes for a "transgenic pig"—a pig with characteristics conveyed to it by two natural parents and a group of scientists. In fact, Roberts is a builder and inventor of tools, and their applications often bear less resemblance to one another than would all the various end products of the welder's art. But *tools* are usually the products of true engineering, of applied science. What is most surprising about Roberts and his shop is that they are at the heart of pure, basic research. Perhaps never before in history has so pure a science been so quickly applied to yield commercially valuble results. The coming of the Gene Age thrust a small group of scientists, most of whom knew each other well but whose names and concepts were little known outside their specialties, into the combined spotlights of Wall Street and the popular media.

Roberts' laboratory, its quiet deceiving, is part of Harvard Medical School, and he was part of the Harvard team under Dr. Mark Ptashne in 1978 that combined a variety of chemical command signals into *E. coli*'s genetic material to force yields in quantities staggeringly larger than had previously been obtainable. For example, of a particular protein that in nature would be cranked out at the rate of 100 molecules per cell, Ptashne's group was able to get 200,000 molecules per cell.

No one has to tell someone in business why yields are important. If your bacterium makes ten times as much insulin as your competitors' at the same cost and quantity, you win and they lose. But to scientists, yields mean something entirely different. They spell the difference between interferon's being an arcane protein whose structure cannot even be guessed and a substance common enough for researchers to study. And with this and other substances, radical yield improvement can transform a laboratory curiosity like interferon into a promising anticancer drug, with several forms now in clinical trials at several U.S. hospitals. Other trials are about to begin to test one form of interferon as a preventive of the common cold.

The industry springing up around genetic engineering is going to take much of its character from these molecular biologists, who, as a group, have a distinct personality. They are single-minded high achievers, capable of great intellectual focus. The top among them are Phi Beta Kappa and/or honors graduates of leading colleges and universities, but they stand out among other kinds of scientists as intensely visual, as imaginative rather than analytical.

Dr. David Baltimore, awarded the Nobel prize in 1975 for discovering a crucial DNA enzyme, recalled in an interview his first interest in biology, sparked in undergraduate honors seminars at Swarthmore College, which had no laboratory. "I wanted to *see* something," he said, a comment wholly in character for this group of scientists. "I could tell that biology was moving into a revolution and the faculty could not. I'd read all the books and papers, but the impetus to see something got me on the trail I've been on ever since."

That trail led the young Baltimore to Cold Spring Harbor, the famous biological research laboratories on Long Island's North Shore, where his requests to see some bacterial viruses led to acquaintance with those who would be his mentors and later his colleagues at M.I.T. That trail would broaden into cancer research, immunology, and the joint founding of Collaborative Genetics (now Collaborative Research), one of the early gene-splicing firms, in Waltham, Massachusetts.

James Watson, mentioned earlier as co-winner of the Nobel prize for first elucidating the structure of DNA, claimed in his memoir *The Double Helix*[1] that he didn't know any biochemistry when he began trying to solve the toughest chemical riddle in biochemistry and genetics. That is probably an exaggeration, but a former Harvard colleague says, "There may be a germ of truth to it. He comes across as exceptionally bright, and he probably has a tremendous *visual* imagination."

Molecular biologists are also practical intellectuals. When they needed a tool to supplement the chemist's standard ultracentrifuge, a high-technology piece of equipment that can cost multiples of $10,000, they devised a system known as gel electrophoresis that makes use of basic laws of motion and molecular weight. The cost of the little plastic gizmo that carries it out: about $30.

Baltimore says of Watson: "I was talking to him about the lab, and it occurred to me that it's part of his genius—and it is genius—to be able to put together a laboratory to successfully carry out advanced biological research; to put together the right people and the physical facilities, and to see the direction things must go, to understand the dynamics of intense, extremely bright people working together."

Baltimore contrasts his colleagues with scientists in other fields this way: "People in biology always had to be somewhat better grounded in reality. We lack theoreticians . . . largely because it is so empirically based a field. You must maintain an involvement over a very long time to be good." And that is why, compared to theoretical physics, he says, "There are very few young people at the top. It takes time to come to grips with all the procedures and techniques and to have them mesh with the theoretical."

Mike Kriegler, a leading researcher in cancer-related viruses, is struggling to deliver an image as he describes the excitement of working with borderline-life microbes too tiny to see, of coming to grips with their compositions and actions, how they work, how they are short-circuited. "You know, I've got it!" he says at last. "It's like Tinkertoys. You're building and rebuilding with Tinkertoys until you understand everything perfectly." Kriegler, speaking at the Fox Chase Institute in suburban Philadelphia, was then on the verge of several key discoveries concerning an even tinier molecular construct called tumor necrosis factor, and he would prove to be an interesting bridge between academe and industry.

Our popular image of scientists at leisure was left us by Einstein: playing classical violin, reading, abhorring physical exercise. But many biochemists are "physical," the kind of people who blow off steam hard rather than easy. When Tom Roberts was Lynn Klotz's graduate student, they played every pickup basketball game they could find, along with others who are still their major colleagues and contacts. Mark Ptashne is known to be a good squash player and top-level Ping-Pong player—as well as a devoted violinist—and Jan Pero is said to be an excellent volleyball player. Vicki Sato describes herself as "nonathletic"—but her major avocation is ballet, and when she resumed dancing as an adult after years of absence, "I danced for hours every day"—as demanding, intense, and physical an activity as any sport.

And these examples are not anomalous. A top yeast geneticist,

Gerry Fink, was a varsity basketball player in college. Herbert Boyer, the founder of Genentech whose teaming with Stanley Cohen turned laboratory gene-splicing into an industry, played football in the Pennsylvania country where it is king. He was voted best athlete by his high school class, but foresaw a career in molecular biology when he did a college report on DNA soon after Watson and Francis Crick unraveled its secrets.

Of course, such characterizations cannot be universally applied. Dr. Walter Gilbert, Nobelist and a founder of Biogen, doesn't think of biologists as more or less physical or practical than other scientists. In fact, he began his career in theoretical physics, which he taught at Harvard for several years before switching to molecular biology. Dr. Shing Chang, a leading expert on the bacterium *B. [Bacillus] subtilis,* also feels that such statements overgeneralize.

Significantly, women make up perhaps half the top researchers in genetic engineering fields, a marked difference from the situation in most other hard sciences and engineering. An officer of one rDNA firm said half its program managers and other doctorate-level staff members are women, as are about half its bachelor-level staff.

Sato said, "It's hard to tell whether women are now more conspicuous just because they've been in biological sciences longer or if there's some other reason that biology was always less sexist" than other sciences. She added that her female colleagues in physics and other departments when she was a faculty member at Harvard faced "a great deal more sexist pressure" than those in the biological sciences.

Sato, director of Cell Biology and Immunology at Biogen, is a close friend of Biogen founder and former chief executive Gilbert, and while at Harvard her office was just down the hall from his—another interesting feature of the "genies." How far the field has come notwithstanding, its pioneers were a small, close-knit group who mastered extremely difficult skills at a time when few outside their field were interested in them. They studied together, competed in adjacent laboratories, collaborated as postdoctoral students. Some date one another, a few are married to each other.

How fast has the field moved? Five years ago, Princeton University's Jacques Fresco pointed out that only a handful of people could splice a *particular* gene into bacteria and get the desired protein in quantity—a "handful" meaning perhaps twenty to fifty scientists in the world. In the summer of 1985, Cold Spring Harbor began teaching a course in recombinant DNA techniques to show high school biology teachers how to carry out true genetic engineering experiments. David Miklos, communications director, who runs the course, explained that the teachers begin with a basic review of molecular biology, and conclude weeks later performing an experiment in which they insert genes of interest into *E. coli,* achieve expression, then grow up enough of the *E. coli* to break out the protein encoded. The aim is to teach high school students these techniques that are revolutionizing science and industry. The following year, says Miklos, "We took the show on the road," taking a laboratory van first to area schools, then to California.

Of course, like Tom Roberts, the other pioneers of genetic engineering have not been idle in the decade since only a handful could splice genes: molecular biologists are among the most intensely competitive people in science. Some executives have questioned whether academic scientists will be able to take the pressure of industrial competition, to slug it out in the "real world." There are a lot of "real world" questions the scientists worry over—but whether they can compete is not among them. Asked about rivalry during her academic career at Harvard, Vicki Sato said, "The competition [in science] is murderous. People right down the hall will be working on the same thing you are and you're trying to beat one another. As soon as one of you publishes, the other hops on—to find a mistake, or to use the results if there are no mistakes. Competition is the name of the game, for results, for grants, for the best academic jobs, and for tenure."

In Horace Freedland Judson's thorough history of the genetic revolution, *The Eighth Day of Creation,*[2] Ptashne recalled months of "absolutely hair-raising" competition with Walter Gilbert, who was working just a few labs away at Harvard. The two were trying to find the "repressor," a chemical that prevents genes from ordering the manufacture of proteins. Ptashne recalled his exhilaration and terror, a feeling of certainty that one of them would be the discoverer, not knowing which of them it would be.

In 1967 each found a different repressor. They both won the race, though Gilbert had a slight edge. The repressor turned out to be of vital importance to molecular biology and to its offspring, genetic engineering.

Gilbert now says the "competitive element between Mark and me has been overstated." But when asked if scientists are not in fact used to intense competition, he laughs and says, "I don't think that will be a problem" as scientists enter the high-tech business.

Gilbert has probably been closer to all sides of the competitive action in genetic engineering than anyone. Chairman of Biogen's science board from the outset, he assumed command of Biogen in early 1982 as chairman of the board and chief executive officer, leaving behind a tenured professorship at Harvard. His three-year term at Biogen was, by all accounts, a stormy one. In December 1984 he returned to his Harvard laboratory.

Vicki Sato, a close personal friend, says Gilbert "never was averse to large questions," and is pursuing a host of entirely new ones as well as developing some of the science he helped to create—the techniques of rapid gene seqencing for which he shared the Nobel prize with Frederick Sanger. Current interests: Neurobiology, the molecular biology of the brain and nervous system.

Still, there are differences in the competition experienced in a university from that in private industry. The race in academic science is to learn new information and get it published as fast as possible. The rush in industry is to produce the most salable product, and that may require keeping results secret until the moment is right. That difference obviously can have major effects on the speed with which information flows and thus on the overall rate of development of new products.

But in one respect, life may be easier in business. In science, the first person to publish almost always gets all the credit, even if the second is not far behind and has done a better job. But in industry, you don't always have to come out with your product first. No one remembers slight time differences. If your product is better, or cheaper, or you market it more skillfully, you can come out very much ahead without being first.

The Race

The genetic revolution began in 1953, when biologist Watson and physicist-turned-biologist Crick, both at Cambridge, got their hands on the best set of X-ray crystallographs of DNA available, taken by Rosalind Franklin. With Crick's knowledge of physics and chemistry and Watson's visual intuitions and "hard thinking" (a colleague's words), they built the first model of deoxyribonucleic acid. The model climaxed an intense race with Cal Tech's Linus Pauling, then considered the best chemist in the world, and others in both Europe and America. It was a race full of clashes of wills, of whimsy, and of bitterness.[3]

Watson sharply criticized Franklin in his memoir for pushing off in wrong directions, even though he would not likely have found the secret without her X-rays, and Watson drew Crick's anger for publishing *The Double Helix,* which his erstwhile partner found undignified and too critical of colleagues. Maurice Wilkins, who shared the Nobel prize with the two, was also sharply critical of the late Franklin, but he said to science historian Judson: "DNA, you know, is Midas's gold. Everyone who touches it goes mad."[4]

If not truly the cause of madness, DNA is surely the driving inspiration of those pursuing the most intimate secrets of life and its controls—a molecule described by more than one scientist as elegant: spare in design and symmetrical, like a Greek temple or a perfect circle. For many years, both Watson and Crick remained at the forefront of the science they helped found. Watson turned Cold Spring Harbor Laboratories from the small, highly respected retreat center it had been into a mecca for scientists in the many allied fields surrounding molecular biology, carrying out cancer research, investigating monoclonal antibodies, examining viruses, training teachers and journalists as well.

Francis Crick was a leader in the race to crack the genetic code and made significant contributions that led to its unraveling years before most scientists expected. Then, ten years ago, he decided to pursue a lifelong interest in the biology of the brain, something completely unrelated to his previous work or even, at that time, to molecular biology. He changed course, set sail in this new direction, and never looked back.

The hot area of biology in the mid-eighties? Neurobiology. But perhaps it should be no surprise that a leader in unraveling the secret of life should now be quietly trying to unravel the secret of consciousness. More of that later (see Chapter 10). We're interested now in learning about the progress of molecular biology from that day in April 1953, when because of Watson and Crick, it took a quantum leap forward.

Most accounts of biology's rapid transformation in just over thirty years cite the Watson-Crick hypothesis as the climax of a heated, worldwide competition—which it was—but leave an impression that the two scientists became instant celebrities among their colleagues. Not so, Crick recalled, in October 1986.

"It wasn't clear when we first published our paper that it was at all important. Many people ignored it. Only a smallish number paid a lot of attention to it. It was thought interesting by some, but it wasn't until years later that it was seen as important. In fact, a lot of people told me they'd disregarded it or thought it was nonsense—and I'm talking now about leading scientists."

The landmark paper is only a half-page long and was reprinted by *Nature* on the thirtieth anniversary of its publication. Compared with most of the dense reports that run in scientific journals, understandable generally only by people within the relevant field and even subspecialty, the Watson-Crick paper is remarkably clear, descriptive, and straightforward, and no term in it would be foreign to college-freshmen—or a good many high-school—biology students.

Crick said that from the point at which he and Watson unraveled the shape of DNA and the critical relationship of its base pairs, he foresaw the development of molecular biology all the way through 1966 "with only one mistake"—a confusion of which kind of RNA was the important "messenger," an error that was corrected by the late 1960s. Perhaps his most important contribution after 1953 was in drawing attention to the tremendous problems involved in cracking the genetic code. It is as easy to read the genetic code (see Chapter 9) as it is to read a telephone book, but in the early sixties, finding the correspondences in the code promised to be as difficult and long a process as deciphering hieroglyphics.

Crick and Sidney Brenner set out to crack the code. In the end, Crick says, his own "real contribution was to draw attention to the problem," lecturing on some of the inherent difficulties and the importance of solving them. "You see, we all did a great deal of thinking about it, but it turned out not to be a theoretical problem at all, but an experimental one."

In other words, many molecular biologists were initially trying to develop a logic to correspondences in the code. There is one kind of logic, for example, in the alphabetical ordering of the telephone book, another kind of logic in the progression of arithmetical or geometrical sequences. As we'll see, it turned out that the only way to figure out what corresponded to what in genetic code was to "go look." There is no logic to the correspondence between one's last name and street number; those devising directories must doggedly accumulate the data, because there's no way to figure it out. Similarly, there's no way to figure out genetic code except experimentally. Once that became clear, the code was cracked within a few years.

Crick and Brenner made a major contribution toward cracking the code when they realized it was a triplet; that is, that each "word" in genetic code was composed of three chemical "letters." But that carries us though the period that Crick "foresaw pretty well" in 1953. His surprises would come a decade later.

Scientists spent the 1960s unraveling the tortuously complex details of how genes "express" themselves to order the manufacture of proteins, then how the strings of molecular beads called amino acids fold themselves into often relatively massive three-dimensional proteins.

To realize how critical the relations between DNA-function and protein-function is, consider the case of sickle-cell anemia, a disease that primarily affects blacks, and kills by reducing or eliminating the ability of the blood to carry oxygen. A *single* incorrect instruction in the DNA chain puts the wrong amino acids in just one link of this massive protein chain. This distorts the molecules' shape. The result is that molecules stick together and cannot pass through capillaries.

If scientists could someday correct that false order in, say, the fertilized egg, the child would never get the disease and might not pass the trait on to his or her children.

Molecular biologists of the 1960s also were scrambling to find the mechanisms that turn genes off and on—the repressors, thought to act like a mask or hand that covers a light until the right chemical conditions remove it and the light can begin signaling.

Beta galactosidase, for example, is an enzyme that breaks down milk sugar (lactose) in digestion. A gene directs the manufacture of this enzyme out of the building blocks floating in the cell, but normally it is quiet because it is masked by a molecule called the lac repressor. When a milk-sugar molecule enters the cell, the lac repressor is pulled away from its normal position on top of the galactosidase gene, the repressor literally combining with the lactose molecule. The galactosidase gene is now free to encode the enzyme that will break down the milk-sugar molecules.

Harvard's Gilbert discovered the lac repressor. Max Perutz told Judson in 1968, "The greatest discovery in molecular biology in 1967, and the one that's received the least publicity, is the genetic repressor—isolated by Walter Gilbert and Benno Muller-Hill at Harvard."

Gilbert won the Nobel prize in 1980—ironically not for the lac repressor but for equally pioneering work in developing techniques for reading the sequences along DNA molecules—ironically, because a colleague notes, "Had Gilbert won it for the repressor, Ptashne might have shared it too."

This colleague describes Gilbert this way: "He is a brilliant, first-rate scientist, with an uncanny ability to know exactly what is going on in his lab and to separate good procedure from bad. At times arrogant, I feel, but absolutely brilliant."

Arrogance is not a foreign word in big-league science. Ptashne is also described as brilliant, with more than a touch of arrogance and a difficulty in working with people. But Ptashne "was awesome in getting things done," the colleague recalls. "He and Tom Maniatis—then a postdoc [postdoctoral fellow]—set out to work on repressors that I figured would take five postdocs a couple of years. I mean, nice idea, but how were they going to carry it out? Nine months later they'd accomplished what they set out to do. Mark is terrific at visualizing an argument as he's hearing it and being able to sift the good from the bad. A lot of conceptual skills."

David Baltimore, meanwhile, made one of the major discoveries concerning the reproduction of cancer viruses. Many are so simple they do not even have DNA, but possess only a slightly simpler, related carrier of genetic information called RNA. Without DNA, how could a virus take over a cell's metabolism to force creation of more viruses? Baltimore discovered an enzyme that the virus "orders up" that *makes* a copy DNA using the simpler RNA as a template. That was one step in a direction biologists did not believe genetic information could move: It was always supposed to flow from DNA outward, never coming back to alter the DNA.

Baltimore now talks with enthusiasm of the work he is doing

in his M.I.T. lab—pure research into cancer. He is a director of M.I.T.'s Whitehead Institute, whose mission is to learn the intricacies of differentiation, the development of organisms from fertilized egg to whole creatures of specialized cells and organs. He points out that his 1975 Nobel prize came for the first experiment he ever did with cancer virus, the first step on what has proved to be a major path for him.

Baltimore and other investigators around the country, for example, now believe that cancer is not always caused by a virus, but is only carried by a virus that has picked up the cancer gene from a cancerous cell. Current concern with discovering cancer genes—several already have been isolated—marks a major change in biologists' thinking. But Baltimore's excitement is over the "pure research," he points out. He does not expect to find a cure for cancer as a result of it. The recent status of cancer-gene research is discussed at the end of this book.

Still, several years ago, losing none of his desire for pure research, Baltimore began to want to do other things as well: practical things, work that would have an application to problems in the real world. Hence, he developed Collaborative Research, a gene-splicing concern.

In research, the early 1970s marked the major milestones in the development of genetic engineering. Scientists knew that each cell in the body contained a molecule of DNA, and that in a given individual, every DNA molecule was virtually identical, repeated billions of times but repeated, presumably, nowhere in the universe outside that individual—unless, of course, he or she had an identical twin. How did the same DNA make a blood cell when it was in the blood, and a brain cell when in the brain? How did the single fertilized egg in the womb differentiate into all the tissues and organs?

Stanford's Paul Berg decided that he would look at the DNA of higher organisms to help answer such questions, as we noted earlier, and both "recombinant DNA" and the controversy over risks were born.

The angry, confused atmosphere that sometimes accompanied the debate at public gatherings is believed responsible for at least some major drug companies initially backing off from recombinant DNA research—enhancing the chances for the smaller gene-splicing firms that dominate the news and, so far, the breakthroughs.

As news reports of the past decade show, major companies now have sunk millions of dollars into genetic engineering, and some have been able to lure leading academic scientists into their own laboratories, but venture capitalists and executives interviewed said the lull in the big companies' interest in the late 1970s was what gave the startup recombinant DNA firms their room to be born.

Not long after Berg's 1973 recombinant DNA experiment, Herbert Boyer and Stanley Cohen met. Their aim was not different than Berg's. The difference was in the proposed technique—a technique that became a technology. Cohen had decided that the best vehicle for recombinant DNA would be the plasmid, the circular ring DNA found only in bacteria. Cohen planned to insert into this plasmid ring the gene that he ultimately wanted to work inside *E. coli.*

In a conversation at a Hawaii convention, Boyer told Cohen he had discovered enzymes that made lighter work of cutting up and rejoining the plasmid ring with its newly inserted gene. These specialized "restriction enzymes" are each designed to cut up very specific DNA sequences—their function in cells is to snip up and thus destroy invading foreign DNA. Other enzymes in Boyer's genetic tool shop, called ligases, would make each end of a cut piece of DNA string attract the other end—in effect, making the ends "sticky" so they could rejoin.

Crick's surprise: "Right in 1953 we foresaw things pretty well through the unraveling of the genetic code. But we did not see the powerful tools which would be developed in the last ten years, the restriction enzymes, the ligases and other enzymes, the gels for separating DNA fragments cleanly."

Crick refers to the slowness of some to realize the importance of the 1953 discovery with no hint of arrogance, as there is no hint of it in any of his conversation. So there is no hint of regret or embarrassment when he contrasts:

"Nobody foresaw this [sudden advent of genetic engineering], not even a year or two ahead. We were all so blind! We had plenty to think about, of course. Then it happened—not one thing but really an accumulation of many discoveries adding up to such remarkable advances—and after '76 everything went more quickly than anyone could ever have expected."

New genes spliced into a ring of DNA were the beginning of recombinant DNA (or rDNA) as a technology, as engineering. A few years later, Boyer joined with venture capitalist Robert Swanson to form Genentech to make such products as insulin and human growth hormone. The race was on to engineer better strains of *E. coli* and learn more about how genes express themselves, and it was at Genentech that the "genies" first discovered "maximization of the gene expression"—that is, overproduction.

Meanwhile, back at Harvard, Ptashne's group was running neck and neck with Genentech. Keith Backman conceived many of the key ideas for maximization, Tom Roberts recalls, while Roberts "worked out most of the early technology." Ptashne later would be the major scientist in Harvard's plan to set up its own genetic engineering firm, a plan in which the university would lend its name and laboratory space in return for being a minority shareholder in the company. That plan was dropped after faculty members loudly protested that the university had no business in commercial enterprise. But under Ptashne's scientific guidance, the Genetics Institute began as a private company, with heavy venture funding and then the contract work most such firms do in their early years.

Roberts, associate professor of pathology at Harvard Medical School, has joined BioTechnica International as a consultant. Backman left M.I.T. to work for this new firm full time. Almost everybody is in somebody's company, and the companies are mostly unions of scientists still in academia with small ambitious firms funded by Wall Streeters, drug houses, and giant energy and chemical firms. They fell together, like a chemical molecule, because the elements were there *and* conditions for the particular combination were right.

And suddenly the praise and criticism that passed regularly among those at the top, as it always does, the living and dying by each other's assessments in a small interwoven group, became the rumor and the speculation of Wall Street. Roberts, modest as he describes his work at Dana Farber Cancer Institute, is praised as one of the country's best genetic engineers, an expert on the molecules that promote gene expression. Asked who *he* thinks is good, Roberts names Genentech's David Goeddel, a major pioneer in overproduction; he's only met him once or twice but knows his work *real* well. "I saw some of his experiments and they were very, very nice."

Top pitchers in the big leagues, and science at America's leading universities is as big as leagues get. Now anybody can clone things. Sure, but anybody can play baseball, too. But not everybody can play in the World Series, and that's a bit like what the craze in the marketplace has been in the ensuing years: Who would come up with the first big "boomer," the billion-dollar or at least multimillion-dollar find that would push a struggling company head-and-shoulders above the rest. As we'll see, the reality turned out more interesting than the forecasts. By the end of 1986, no one had hit that "boomer," but discoveries, products, and revenues were pouring in from an astonishing number of companies in an unforeseen variety of diagnostic and therapeutic areas.

The sports analogy is helpful in another way in assessing this new, uncertain industry: in high technology firms especially, success or failure depends almost entirely on people. One scientist says, "High tech industries are people-intensive. We don't have a mineral to mine, or a forest to harvest. What we've got is the product of people's minds. How good they are, how well they work together: those are the best things that determine the outcome."

Roberts says Genentech is "a model of one type of company in our field. It *is* its employees, really. You can follow their pattern in this field: if you're interested in a particular [university] lab's process, hire the best postdocs out of the lab for your company." In postdoctoral work, scientists gain valuable research experience before striking out to run their own labs. Among Genentech's innovations in the marketplace: 25 percent of the company's shares are held by employees. Employee stock options are not new, but few companies have such high employee ownership.

DNA Science, the E. F. Hutton creation discussed in Chapter 1, was another type, aiming to back university research in the universities where the best work was being done.

In an interview in July 1981, DNA Science president E. Russell Eggers called the setup "the most interesting cocktail of money in the country," and he was no neophyte enthusiast. Nevertheless, the plan collapsed. Hutton continued to operate a stripped-down version of DNA Science, but without the features that made its original plan unique. It was to have been a marriage of university and Wall Street, and the failure indicates some of the strains of these new partnerships.

But why the straining to create such a partnership? Why all these small companies working for large firms on contract, with other firms in effect trying to broker deals between university labs and commercial gene-splicers? A simpler scenario would have Eli Lilly, Monsanto, Standard Oil, and other giants simply hiring the country's best cloners at any price and accommodating them in corporate labs. To some degree that has occurred, but it is still the exception rather than the rule for one of the country's top scientists to be part of a big company, and then it is usually at the vice president level. Traditionally, these giant concerns have had an even larger image problem with scientists.

In fact, until the past few years, leading university scientists typically referred to work done by larger firms as third-rate science—even if it represented first-rate technology and marketing. Most of the major breakthroughs in pharmaceuticals in recent decades had been achieved in university laboratories associated with hospitals, then sold to big drug houses, where they were transformed into marketable products.

The best students in top university laboratories are still urged into postdoctoral research and from there into academic positions. A young Princeton professor was overheard talking to an undergraduate who'd just announced she'd decided *not* to go to medical school—even though she was sure she would get in—but to get her Ph.D. and do research instead. "Attaway!" he said, hugging her in rough camaraderie. "I knew you wouldn't sell out."

If medical school can be construed as aiming for less than the top, where does that leave even the best of the drug houses in the fierce competition for these best and brightest?

Whitehead director Baltimore says most of his graduates move into academia, and that appears to be the first step for most top graduates and postdoctoral students, and most academic scientists interviewed agreed. But there has been a major if subtle change over the past few years. Some top graduates are moving from academe into the young recombinant DNA firms, and some top senior scientists are moving to key science policy positions at the more foreward-looking major drug companies, as well as the smaller firms. This shift, though small in numbers, represents a sharp change in thinking by scientists in very few years.

If nonacademic science has gone from "third rate" to often-acceptable, that has meant an even more dramatic difference to those graduates who are just under the very top. As several scientists pointed out, a few years ago if a young graduate did not get an academic job, the outlook was very bleak for salary, research opportunity, and acceptance. Now, as Cetus's Ronald Cape told us, "A young graduate in molecular biology can have a justifiable expectation of making a good income and of doing interesting work, regardless of which career path he or she chooses."

This new attitude evolved in the small companies, beginning with Genentech and the host which have followed or which have adapted their earlier research plans to focus on the new biotechnology. They emerged, literally weekly through the first half of this decade, to bridge the gap between university and industrial laboratories. Some small firms were founded by leading molecular biologists, others hired them with major equity shares and, equally important, real authority. So many Nobelists have become involved that their presence on boards of directors and science advisory boards has sparked jokes in both business and science. Science advisory boards are being headed by people who have made the science, and many sit as company directors while remaining at their university research posts. Genentech's Boyer is still a full-time professor at UC San Francisco, for example.

Walter Gilbert pointed out that Biogen began as a group of scientists with venture-capital backing, not the other way around. It was the scientists who recognized that many of their discoveries were ready for development and who knew how that development ought to proceed, he says.

David Baltimore also sees the emergence of the companies as far more than a technological linkup. In fact, he said, long ago he saw Collaborative Research as an important part of his work as a scientist, albeit one totally separate from his research. "Things done in basic research have an impact on practical problems, but no one knows what that impact is, it's completely unpredictable. Naturally, you have the desire to utilize research skills in solving practical problems, in bringing real change to life. You've got to somehow mesh basic research and application, so you have to face the needs." The question is how?

"You *could* do it in a university, but you would scramble for grants that shouldn't be there for work that is going to be profitable. The nonprofit sector doesn't support such work in a capitalist society. When you're working on something with those practical, commercial applications, then you ought to operate as a commercial enterprise and not on a grant of public funds."

Baltimore is particularly proud of the work Collaborative Research has done in recent years in identifying gene sequences that are known as "genetic markers." Simply put, these are telltale combinations of base pairs that always show up in the area of particular genes. For that reason, they can be used to locate mutated genes that lead to the genetically inherited diseases that have ravaged humans since ancient times, such diseases as multiple sclerosis, muscular dystrophy, Huntington's chorea, and cystic fibrosis. Collaborative was recently cited as having identified more of these markers than any other laboratory—and this is basic research with a practical aim. The firm hopes to develop diagnostic probes that will identify those at risk for developing these diseases or for passing them on to their children. It has already developed several such tests, and recently, Baltimore says, located the gene that causes cystic fibrosis.

Baltimore is chair of Collaborative Research's science board and is a member of the board of directors. Yet his own research at the Whitehead is completely remote from this field. He is studying polio virus in an attempt to understand the molecular biology of the immune system. In fact, the discoverer of the now "hot" reverse transcriptase says he does very little work with this revolutionary enzyme any more, having moved in new directions.

Of the controversy over pure research versus profit, Gilbert said: "There's been a completely erroneous attitude around that we're seeing basic research that has become of sudden, extreme value. Not true. Interferon is a perfect example. That basic research into interferon probably costs some fraction of a million dollars altogether. The cost of the applied research in the *companies* trying to make it will be on the order of $20 million to $50 million. What happened in the last decade was rather the realization that this research *would* be of great commercial value" many millions of investment dollars down the road, "not that it *was*."

On paper, the commercial value is working out in stunning figures. After Genentech's landmark public stock offering in October 1980, Boyer was worth about $40 million, a figure that was loudly publicized both in jubilation and consternation.

The effect on academe: electric. Good scientists have been willing to join up with the small firms because they retained their ability to do basic research and, most importantly, to publish. But certainly the financial rewards have not gone unnoticed. Says one molecular biologist now with a small rDNA firm: "Look at me, I'm 40 years old, I've spent my whole life in academia and I'm at Harvard, which is the top. And what do I get for it? To live like a poor student."

A small science community suddenly finds itself thrust into the limelight—and into a position to change the world. The combination is thrilling and unsettling.

New Game at Old Ivy

Vicki Sato recalls the excitement and turbulence of 1981, when she and her colleagues knew they were making decisions at a pivotal time for the genetic engineering industry. But she says: "Moving from academe into biotechnology we didn't ask, How will *I* be different? Biotechnology's first incarnation was in pharmaceuticals, yet we never looked to the ethical-drug area as a guide to how we would develop. It was a point of pride that we were going to be different. Yet I think the truth is that a lot of the character of our industry derives from the character of the pharmaceutical business. It has left its imprint on us, and it's funny to think of that way."

Crossing a grassy square in a quadrangle of old brick buildings in mid-1981, you suddenly look at the life-size bronze rhinoceroses that mark the entrance to Harvard Biological Laboratories. Inside, the old labs offer no testimonial to the importance of their scientists or the power of their discoveries. Walter Gilbert's lab is here—he has not yet made the momentous decision to take direct control of Biogen as CEO. Down the hall is Mark Ptashne's, and it is the summer after the founding of the Genetics Institute, which also would become a leader in recombinant DNA. Gilbert and Ptashne are said not to be on good terms. "You don't invite them to the same parties," a scientist says.

Up and down the halls, the talk is of companies. His company and her company. He consults for this one, she is starting that. Vicki Sato, laughing at the business boom, is no exception. "Everybody's got a company, and I mean everybody." At this time, hers is Angenics, a firm she helped found to put monoclonal antibody technology into the marketplace, and for which she would soon leave her associate professorship. "When we get together for coffee, the conversation is always about what firm somebody has started or joined."

Uncommon conversation for basic-research scientists, for these academic laboratories have been their native soil throughout the evolution of science. Because of Sato and Gilbert's personal friendship, starting up Angenics was easier for his ready advice. But what about restrictions on the flow of scientific information brought about by their allegiances to two different companies? "I found it to be the opposite," she said. "Wally [Gilbert] and I talk more about science now than we used to. We talk about the work we're doing at our companies as well as the work in our labs. Sure, maybe at some point in the conversation we say something general or vague in order not too give away too much—but that's no different than when two scientists are competing in the same academic field."

No one expected to give away trade secrets, and in any case, molecular biologists had never been inclined to give away the farm. At that stage in genetic engineering's development, the style of competition practiced in academia—the race to be out first, the race to use the good work of competitors to surpass them—was among the healthiest signs for the nascent industry.

And there were further signs that the scientists were prepared to commit themselves to playing by the rules of a new game while not abandoning those of their profession. Sato commented, "We're trying to build a business. We're not trying to make a haven for basic science in the absence of good business."

The intimacy of this small group of scientists who found themselves on the brink of revolutionizing older industries would seem to help genetic engineering survive as an industry. In the event of the severe shake-out of companies, the bonds of scientific camaraderies might hold. As we'll see, that shake-out has not occurred and probably will not, not in the shape and with the disastrous effects as were then being widely forecast. But the comments and the concerns hold. In those days, still so recent, scientists themselves foresaw a strange new world and tried to read their places in it.

Not surprisingly in this small group, conversations are sprinkled with asides on who's married to whom. Tom Roberts and wife, Gail Lauer, both pioneered overproduction, or maximizing gene expression; she specializes in the control regions of genes and left Harvard to carry on her work at BioTechnica.

Richard Losick, Harvard professor of biology, is married to Janice Pero, a former associate professor, and both specialize in the genetic mechanisms of *Bacillus subtilis,* a bacterial "factory" that operates differently from *E. coli,* with potential industrial advantages.

Maybe business is the new shop talk in the lab, Losick and Pero say, but it took time for them to become accustomed to the sudden "glamour" foisted on their pet bacterium, just a couple of years ago known in America only to industrial fermenters and academic biologists. Pero says, "We've spent our careers studying the control regions of *B. subtilis,* so it's always been important to *us*. But neither of us foresaw the possibility that what we were doing would have commercial value for decades."

Losick adds, "We were doing pure research, because it interested us. All of sudden my mother is asking me what stock she should buy. I think it's crazy, the way the industry has taken off."

But Losick and Pero, though part of the bio-boom and looking forward to seeing their research benefit the public, were apprehensive about public perceptions that this technology would begin paying off handsomely right away. "If I were going to put my money in these startup companies," Losick said, "first, I would want to be very young; second, I would not expect to see a lot of return for many, many years. And with those provisos, yes, you might make a lot of money in this area in twenty years, or you might lose it."

But for the scientists, there were other considerations in their own involvement. Pero pointed out that because the small firms were carrying the technology to market, she would be able to go into industry and remain in Cambridge as a senior scientist holding equity, rather than have to join a huge drug firm that allows little scientific freedom and requires relocation away from academic centers. That was a major draw for her and many like her, allowing for a transition before there was a softening in the attitude toward the larger firms. Pero has been senior scientist with BioTechnica for several years, still working with *B. subtilis*. Losick, meanwhile, has become chairman of Cellular and Developmental Biology at Harvard.

Tom Roberts pointed out that of job offers he has had from rDNA firms, "The largest was three to four times the salary I make at Harvard." Luckily, he said, wife Gail Lauer decided to become a full-time scientist for an rDNA firm, and he will consult. "Otherwise I'm not sure I could've withstood the pressure."

These early conversations, like the 1981 attitudes about the ease of transfer of technology into the marketplace that they were encountering, seem long ago, of another world. What followed for the "genies" in business was what had preceded in science: work, and more work, only now with an added element of turmoil not associated with basic research—the musical chairs of the marketplace as scientists moved into business, sometimes moved back again, and everyone seemed to be changing places at the same time.

Vicki Sato and Lew Cantley were married and have a daughter, Mariko, now three years old. After Sato's initial venture in Angenics, she joined Biogen to set up its department of cell biology and immunology, which she still heads. Cantley is a full professor in the Department of Physiology at Tufts Medical School, a basic researcher into the molecular basis of cancer, and by the coincidence of their acquaintance found himself collaborating with Tom Roberts on key experiments into the nature of cancer-causing genes. Sato's friend and mentor, Walter Gilbert, is back here at Harvard Biological Laboratories again, a building away from former competitor Ptashne, who is in the new biology labs.

Sato now says, "The switch to pharmaceuticals was the right one for me." But she notes that the drug industry in many ways was a strange one for genetic engineers to enter in force first. "This was the first instance in which high technology chose to apply itself to the industry with the longest time lag of development." From the perspective of giant drug firms, the move was beneficial, she says. "The big companies have had the luxury of investing money, then sitting back and waiting, then increasing their investment if something looked likely."

Sato's question is: "Will that serve as a source of rational drugs, or as a reservoir of biology to allow you to understand different mechanisms? Or will biotechnology become a service industry of pharmaceutical houses?"

Sato noted that plant and other genetic engineering was not at issue here, but many startup firms faced with the end of venture money and no near prospect of survival revenues went into joint ventures with drug companies. Therefore, "you need to pay attention to the philosophy of the investing industry."

There turned out to be more unity in science and business than expected. If scientists appear driven in their research, their own words imply that the motivation is more a pull than a push. There is an urge, a quest for answers. Practical things are not necessarily part of scientists' quests, but there seems no contradiction in finding them in a scientist's repertoire. Practicality, on the other hand, is a logical necessity for survival on Wall Street, but that questing nature is often found there, too, where the favorite old slogan preaches: "No guts, no glory."

John Hunt, among the most unusual of the genies, himself offers a telling comment on the willingness of many people in high places to commit themselves. "We're right on the frontier," he said in his office at Princeton's renowned Institute for Advanced Study, where he was associate director. "The world after this is going to be a different world." Though not a scientist, he has spent his life among scientists. Author of two novels, he was nominated for a National Book Award for his *Generations of Men.*[5] A "kid out of small-town Oklahoma," he rose to become executive vice president of the Salk Institute in California, vice president of the Aspen Institute in New York, and vice president of the University of Pennsylvania. He spent twelve years in Paris and was director of operations for the oldest French foundation, helping to establish and run the Center for the Science of Man there.

He then went to the Institute for Advanced Study, leaving in early 1982 to become chairman and chief executive officer of BioTechnica International, another gene-splicer.

"The changes that will be wrought by the people who have made this leap, who are making it now, will not only transform society," Hunt said. "They will test society at its very center, at the definitions of what life is and what makes it valuable, of what is human, and what is permissible. That's why this is so important."

NOTES

[1]James Watson, *The Double Helix* (New York: Atheneum, 1968), is his much-acclaimed memoir of the race to discover DNA's structure. Reprinted with permission. For readers with a background in biochemisty, James Watson, *Molecular Biology of the Gene,* 3rd ed. (Menlo Park, Calif.: W. A. Benjamin, 1976), is considered the classic textbook on the subject.

[2]Horace Freedland Judson, *The Eighth Day of Creation: Makers of the Revolution in Biology* (New York: Simon and Schuster, 1979), is an excellent and readable chronicle of the recent history of genetics, for those interested in an in-depth account of the people and events involved. Copyright © 1979 by Horace Freedland Judson. Reprinted by permission of Simon & Schuster, a division of Gulf & Western Corporation.

[3]*Double Helix, op. cit.*

[4]*Eighth Day of Creation, op. cit.*, p. 264.

[5]John Hunt, *Generations of Men* (Boston: Atlantic, Little, Brown, 1956).

GENETIC ENGINEERING: THE SCIENCE II

4

IN ORDER to understand the wealth of procedures and tools genetic engineers have at their disposal, we should look first at how a natural organism controls the amounts of proteins it makes at various times in its life. *E. coli*'s daily diet may vary greatly, as does our own, because this bacterium can live everywhere from streams to animal intestines. In one place *E. coli* cells may be surrounded by glucose, in another a quite different food substance. Such different foods call for different enzymes to digest them into needed cellular components; on the other hand, if the cells are temporarily in a "starvation" environment, they have no need to produce any digestive enzymes.

To conserve energy, cells try to limit the making of enzymes and other proteins to times when they are needed. Further, some proteins will be needed in far larger quantities than others. Proteins involved in the construction of large cellular components, such as ribosomes, would be required in huge amounts compared to an enzyme needed only for an occasional chemical reaction in the cell. Such observations lead to some of the central questions in genetic engineering: How does the cell control how much of each protein it makes? And if we can find out, can we trick the cell into using these controls into making large quantities of *our* protein?

Molecular biologists already know a great deal about the control regions of *E. coli,* and they know that a lot of the control—but not all—takes place at the DNA level; that means it can be interfered with easily by genetic engineers. For now, we focus exclusively on control of protein production at that DNA level. Consider the fact that the more messenger RNA copies of a given gene-region are made, the more copies will be bound to the ribosomes, making that much more protein. Then how does the cell control the number of mRNA's made? By instructions coded in the *control region,* a portion of the DNA that sits just in front of the gene region. The control region contains information in that same four-letter (i.e., four-base) alphabet, which determines how often mRNA copies of the single gene are made. Figure 5 is a highly schematized sequence of steps from DNA to protein. We've pictured the DNA as a single horizontal line and have not articulated the sequence of bases, even though that determines the instructions.

Through the schematic microscope of Figure 5 we can see a gene (G), control region (C), and a ribosome binding site (RB), whose functions will be clearer shortly. Bound to the control region of the DNA is a molecule of RNA polymerase (the black rectangle in Figure 5a), the enzyme that makes the mRNA copy. How many copies of the mRNA will be made depends on how strongly it binds to the control region; stronger binding means more copies, because more RNA polymerase molecules will bind and go into action.

After binding to the control region, the RNA polymerase travels along the DNA molecule toward the gene region, as the horizontal arrow shows, making an mRNA copy of the DNA as it moves along. The copying ends when the RNA polymerase encounters a "stop" signal on the DNA sequence after copying the gene (Fig. 5b). Notice that the copying begins before the polymerase actually reaches the gene. That's because the first thing that must be transcribed is the site on the new mRNA that will bind to the ribosome, so the ribosomal machinery can read its information and make the protein molecules called for.

The next step (Fig. 5c) is critical. The mRNA binds to the ribosome at a particular spot *complementary* to the mRNA's own ribosome binding site. Without this very specific binding, no protein will be made. Now bound, the mRNA moves along the ribosome (Fig. 5d). As it does so, the sequence along the mRNA is read and "translated"; the ribosome catalyzes the assembly of the right sequence of amino acids into a polypeptide protein chain, aided by certain adaptor or translator molecules called tRNA, to be discussed later. As the amino acid beads are strung together at the ribosome, they fold into a globular structure, as beads on a

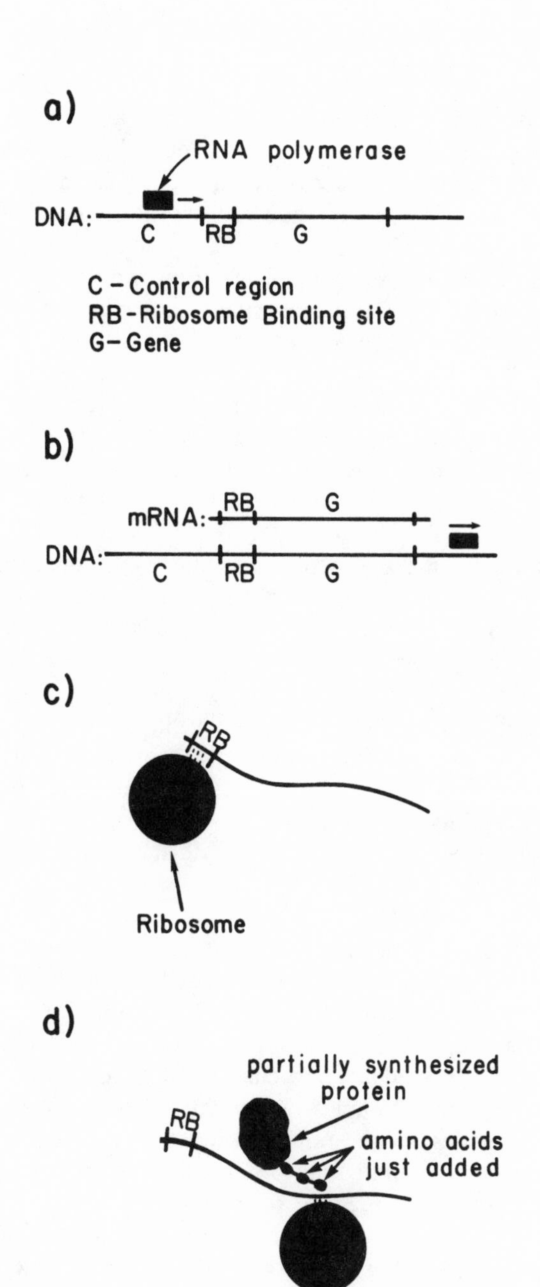

Figure 5. The major events in the transcription and translation of a gene.

string do if piled or rolled into a ball, and that is the completed protein molecule. You can also see a partially finished protein molecule, folded into globular form at the left as amino acids are being attached to the end of the chain on the right. We'll discuss protein synthesis in greater detail in Chapter 9.

Before rolling up the magnification on the schematic microscope, we need to look at protein manufacture in more detail and review what we've discussed so far.

1. The RNA polymerase binds to the control region in front of the gene. Whether the polymerase binds at all to this region and how strongly it binds depends on the base sequence of the DNA; in turn, the strength of the binding determines the rate of mRNA production.
2. The RNA polymerase encounters a start signal and begins copying the DNA as a messenger RNA until it reaches a stop signal telling it to quit. Both start and stop signals are encoded in the DNA base sequence.
3. The mRNA binds to the ribosomes, and another start signal—this one encoded in the mRNA—tells the ribosomes where to begin reading to make the protein encoded by the gene. At the end of the gene, after all the protein's amino acids have been assembled and linked according to the instructions on the mRNA, another signal is hit saying, "Terminate the amino acid chain": protein complete.

Now our earlier assertion should be clearer: all the instructions for making a protein are somehow encoded in the base sequence of the DNA and consequently in its mRNA copy. If we could just increase the magnification of our microscope so we could see how the bases were ordered in a particular DNA gene region—whether ATG or GGC, for example—we ought to be able to read in the language of genes. With enough samples of reading in this language, maybe we could come to recognize the control sequences for binding RNA polymerase, start and stop signals, and so on. In a nutshell, that is how the history of this area of molecular biology proceeded. The techniques used to determine the base sequence of DNA were chemical rather than microscopic, but they accomplished the same goals. They allowed molecular biologists to read DNA sequences that spell command sequences for protein production or that correspond at the end of the production line to known proteins. This is what we mean by "sequencing DNA": learning the order of bases, which allows us to relate that order to commands or to protein segments.

The section of DNA control region where RNA polymerase binds and begins moving "downstream" toward the gene is known as the *promoter,* one of the control region's major components. The genetic engineer's bag of tools includes "strong promoters," so-called because polymerase binds strongly to them, and this efficiently makes an mRNA copy. In weakly bonded promoters, polymerase could diffuse away from the DNA before beginning a copy. We are beginning to understand what it takes to have a strong ribosome binding site on the mRNA, too. Using techniques for splicing DNA pieces, genetic engineers can build DNA pieces containing strong promoters and strong ribosome binding sites spliced in front of the gene for the desired protein. Finally, they can insert this "recombined" DNA into a host organism, which will make the protein in large quantities. This is one of the major tasks of a genetic engineer.

There is also another kind of "control" we need to explain. We mentioned that *E. coli* may find itself at some time needing none of a particular protein, then later needing a lot of it. Strong and weak promoters do not provide this kind of control. *E. coli* can regulate the amounts of *different* kinds of proteins made by having a strong promoter before the gene of one, a weak promoter ahead of the gene for another. But clearly this method does not allow for variations in the quantity of a *particular* protein at different times. These latter controls are known to exist and can be thought of as molecular switches, turning on mRNA production at one time, cutting it off at another. The switches physically block RNA polymerase from binding to the control region and from moving downstream from the promoter to the gene "start signal" by binding to sequences on or near the downstream end of the promoter region. These blocks are known as *repressors,* and those sequences between the beginning of the promoter and start signal where they latch on are known as *operator* sequences. *How* they operate will be part of our deeper look in Chapter 9.

The Genetic Engineer's Tasks

Let's look at the six phases of the genetic engineer's work in some detail; along the way a lot of the phrases used in newspaper and magazine accounts of discoveries will become clearer. (1) Obtain the gene or genes for a desired protein or proteins; the usual methods are called "shotgunning," "copy-DNA cloning," or "DNA chemical synthesis." Which procedure is used depends on the organism from which the gene will be taken and the number of base pairs that make up the gene. (2) Then, build a plasmid or virus DNA to be the gene's carrier or "vector"; it must contain the right control sequences (promoter, operator, ribosome binding site) for the host organism that will be making the protein. (3) Insert the gene obtained in (1) downstream from the control sequences in the vector. How far the gene is from the control sequence is also important. With the right control sequence and distance, the protein will be made in large amounts in the host. (4) Now, insert the rebuilt vector into the host, which will treat the new DNA as its own. (5) Clone a colony from the redesigned bacterium, a fairly simple and time-tested practice. (6) Finally, carry out further genetic engineering to counter unexpected problems, to increase protein production, and to help fermentation engineers scale up the growth of the organism to profitable commercial levels.

Take a look at the first step. Obtaining a protein's gene is usually easy in the case of bacteria or viruses, but often is very difficult with higher animals and plants. Shotgunning is the most frequent method of getting the gene. Here, *all* the DNA from an organism containing the sought-after gene is cut up along with thousands of others. The DNA is snipped into thousands of pieces just larger than gene size, using special enzymes called *restriction enzymes* that cut DNA only at specific sequence sites.

Next, *all* of the thousands of pieces are spliced into plasmid or virus DNA vectors; the vectors must be ones that can be carried in a growing *E. coli* culture. The splicing is done with an enzyme called *ligase.* The opposite of a restriction enzyme, ligase can bond together pieces of DNA that have been cut; it is the "sticky-end" enzyme of Chapter 3. These procedures are directly analogous to a film editor's cutting and splicing. The discovery of the restriction and ligase enzymes made recombinant DNA technology possible.

If the splicing is done right, at least small amounts of protein will be made.

Look at shotgunning through the schematic microscope (Fig. 6). At the upper left are four identical *E. coli* cells, showing only the plasmid DNA in each, labeled P. The single hash mark on each plasmid represents the specific sequence recognized and cleaved by the restriction enzyme chosen for this shotgun. At the upper right is a single cell from the organism that naturally manufactures the protein we're after, showing only the cell's chromosomal DNA. The hash marks on the chromosomal DNA represent the four cleavage sites recognized by this same hypothetical restriction enzyme. After snipping open the DNA with this enzyme, we'll have pieces A, B, C, and D, and it happens that piece A contains the gene we're after, illustrated as a thicker line of DNA.

Our aim: to splice the piece of chromosomal-DNA containing gene A into the plasmid DNA of one of the four *E. coli,* then reinsert the DNA plasmid with gene A *back* into the *E. coli,* and grow up a culture of that *E. coli* to get large quantities of the gene.

The first step is to get both the plasmid DNA from the *E. coli* cells and the chromosomal DNA from the donor organism: there are routine chemical procedures to break both free. Next, all the DNA is snipped with the restriction enzyme, resulting in plasmid DNA with single breaks in them and four pieces of chromosomal DNA from the donor—one containing the desired gene. This whole procedure takes only a few days.

Next, we combine the restricted DNA from the plasmid and that from the organism, splicing them together with the ligase enzyme. This nets four plasmids, each containing a different piece of foreign DNA. All the plasmids now are reinserted into *E. coli.*

You might wonder why the donor gene is not inserted directly into *E. coli*'s chromosomal DNA—simply "stitched" into a break at the proper place in the chromosomal sequence. The main reason is that it is much more difficult to do this stitching inside a cell, compared to outside the cell on a plasmid, and plasmids already have ways of working their way into cells, where they generally express themselves independently of the cell's chromosomal DNA. The mechanism exists because plasmids are naturally exchanged among bacterial cells. Also, plasmids are generally *small* pieces of DNA, on the order of several thousand bases, whereas chromosomal DNA is in the millions of bases. Small pieces are easier to

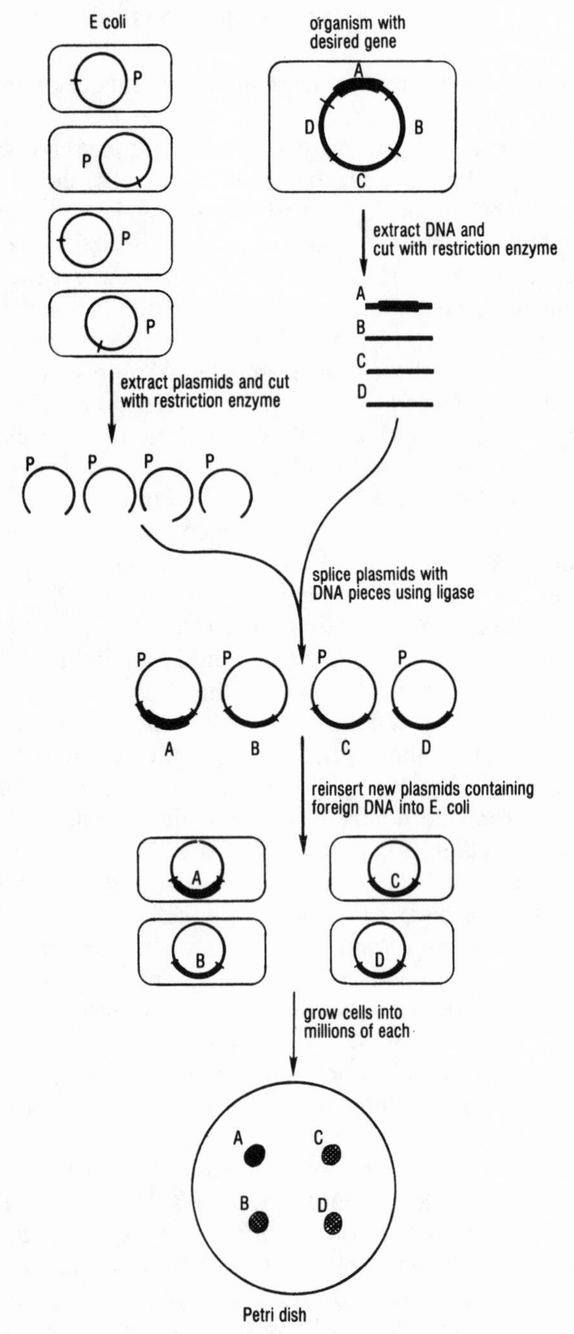

Figure 6. The procedure for "shotgun" cloning of a gene.

work with and analyze for results, making plasmids a good vector for putting foreign DNA into *E. coli.*

The DNA of certain bacterial viruses is also useful, however; these are viruses that attack bacteria, just as some viruses attack humans and other plants. Remember that in its normal mode, a virus attacks by inserting its DNA into a bacterium. Sometimes the viral DNA divides along with the infected cell. In this sense, it resembles a plasmid, although it doesn't usually exist as a little circle of DNA inside cells, but is often spliced right into the main, chromosomal DNA. Viral DNA falls into the same size range as many plasmids, so genetic engineers will sometimes use it as a vector. Looking back at Figure 6, note that the vector must be reinserted into the *E. coli* cells, whether it is plasmid or virus. In the lower center, four *E. coli* cells each contain a plasmid bearing a different piece of the spliced foreign DNA.

Two steps remain to get large quantities of gene A: separate the four *E. coli* cells and grow up millions of copies of *each,* then pick out the copies that have gene A. Separating and growing copies of each *E. coli* cell requires only that classical biological procedure, cloning. Ordinarily, although each bacterium in a group would give rise to its own daughter generations, the bacteria would grow all mixed together; we want them to grow up separately, as cloned colonies. In cloning, we dilute the *E. coli* in solvent to such an extent that there is about a centimeter between any two cells, then we lay the cells down on a gelatinlike substance called *agar.* As with gelatin, agar is solid but contains large amounts of immobilized water; thus the cells will have water but won't be able to swim and mix, and will grow on nutrients in the agar. In time, what was one cell will become a visible, usually white spot of several million cells on the surface of the agar. The bottom of Figure 6 shows a petri dish containing four spotlike clones; one of them contains only cells bearing gene A, one contains only B, etc. Which is which? We could extract DNA from samples in all four colonies and sequence it until we found gene A, but that would be a Herculean task. Our schematic microscope shows only four pieces of DNA in fragments; a typical experiment would have a thousand or more. Even worse, from that huge collection of DNA we would be hunting for one tiny gene region, a needle in a haystack. Sequencing just one piece might take weeks, so this project could last decades.

But after all, the gene we're looking for expresses itself as a protein. Why not find the gene by locating the protein in the only colony that will be making it? Often simple chemical tests will color the colony containing the right protein. In our experiment, the clone containing the desired gene is black instead of cross-hatched. For example, there is such a color assay for the protein beta-galactosidase. If *E. coli* cells contain it, the assay will turn the entire colony blue instead of white—a difference that makes finding the right clone among thousands a snap.

Earlier, we mentioned that inserting a foreign gene into a bacterium is easy compared with getting the gene to express its protein, and clearly that is what we need to do to use these assays. Getting *E. coli* to produce the protein encoded by a gene requires the genetic engineer to do some advance work of the kind we've just described. He or she builds a plasmid containing a single restriction site at a point slightly *downstream* from a promoter and ribosome binding site, because they must be read out by the RNA polymerase first. The plasmids shown in Figure 6 are not natural ones; they represent artificial plasmids built up over the past few years by genetic engineers, each with different promoters, ribosome binding sites, and sites for many different restriction enzymes. Each of many different restriction enzymes recognizes a different four- to eight-base sequence. The genetic engineer pulls out one of these special plasmids, depending on the shotgun experiment planned, then goes to work to break out chromosomal gene fragments. This is just one of the procedures used to clone and identify a particular gene.

"Shotgun" was coined to describe the random nature of blowing an assortment of DNA bits into a group of plasmids and then shooting the plasmids back into *E. coli.* Doing all that and cloning colonies is not difficult; the steps are routine and the procedure takes from a few weeks to a few months. Most often the hard part is finding a simple assay to *locate* the clone containing the desired gene.

In the absence of a color assay, one might try to identify the protein by its function. For example, since interferon protects against animal viruses, animal cells might be injected with serums of each colony to locate the one which provides immunity against virus. You can do such a test with thousands of small colonies, but it is tedious work.

Another method of identifying proteins is to use fluorescent or radioactive antibodies. The right antibodies would bind only to the sought-after protein, and their label could be easily read by laboratory instruments. But obtaining specific antibodies can present major difficulties in itself, so this method may not be usable. In some other cases, large quantities of messenger RNA for the gene being hunted can be obtained. Similarly labeled, the mRNA will bind to the DNA on the right gene region and can be detected with instruments.

In some cases, identifying proteins after shotgunning could take years instead of months, and that is certainly something for a research director of a company to consider before contracting with a genetic engineering firm to do a protein project.

Often the major advantage in shotgunning is that by the time the proper clone *is* detected, the colony is already making the protein; the genetic engineer may only have to fiddle around with the exact placement of the gene with respect to the promoter in order to achieve overproduction of the protein. Once the gene is identified, it can be snipped out, placed in any properly "tuned up" vector, and inserted in a final host for commercial production. At this point, we've covered all but the last part of the genetic engineer's work, but that may be the most difficult of all—preparing for commercial production.

This last phase might require radically increasing the amount of protein produced, controlling that production through the use of the proper promoter and operator sequences, finding a better host organism and engineering *it* to do the job, and adapting the whole procedure to large-scale industrial fermentation. One wonders if it is sometimes not "deliberate oversight" when firms announce major breakthroughs, even though they've failed to achieve a high level of protein expression in an organism suitable for industrial or pharmaceutical fermentations.

We also noted that obtaining genes from animal or plant cells can be very difficult, taking many years in some cases. It's easiest to get genes for small proteins of, say, fifty amino acids or less, compared to large proteins that can possess upwards of several hundred amino acids. Hormones are a potentially interesting class of protein, and usually small. They dramatically affect many bodily processes: absence of human growth hormone, for example, causes dwarfism, and absence of insulin can kill.

Genes for small proteins such as hormones can be chemically synthesized. Instead of pulling the gene from the chromosomal DNA, it is synthesized from the four bases—A, T, G, C—which are commercially available in a chemically active state and are not expensive. This synthesis can be carried out manually or by using one of the highly publicized gene-machines.

Making synthetic genes for 50-amino-acid proteins requires only that 150 bases be strung together in the right order. Such a synthesis, which would have taken a few months just three years ago, now would take one scientist about fifty working hours. Even more remarkable progress has been made in the maximum base size of genes that can be synthesized, and a great deal of this work has even been automated.

The relatively easy and reliable shotgunning technique does not work in obtaining most genes of higher plants and animals, and at the moment so-called copy-DNA cloning, discussed in Part III of the science section (Chapter 9), is the only way to obtain genes for these larger proteins. The reason is the existence of a strange trait in these higher organisms: interspersed among the genetic information coding for protein are gene regions that can only be described as gibberish. Organisms that contain these "nonsense DNA" regions also have means of recognizing them, and they are thus eliminated from the mRNA before it reaches the ribosomes to begin protein manufacture. Bacterial genes don't contain any gibberish; therefore they have no way of eliminating it.

Clone an animal or plant cell gene into a bacterium, and you get a protein whose chain is interspersed with incorrect amino acids coded for by the gibberish regions. The protein would likely be nonfunctional. Thus, cDNA cloning requires an editing process; the end result is a piece of DNA with the gibberish excised. The procedure is very complex in almost all respects, and we'll just highlight its major points for now. First, the genetic engineers must pick up on the mRNA after the point at which its gibberish has been removed. They then use an enzyme known as reverse transcriptase to make a DNA copy *from* the mRNA. Notice that this is exactly the reverse of RNA polymerase, which makes an RNA copy from a DNA region. As mentioned in Chapter 3, David Baltimore's Nobel prize was for, among other pioneering work, the discovery of reverse transcriptase.

The complexities of cDNA cloning are many. It's often very difficult to obtain the necessary mRNA; making the cDNA copy is not easy; and finally, further preparation of the cDNA copy is needed before it can be inserted into a plasmid vector. The whole procedure could take years for a particularly hard-to-find or particularly large mRNA. In the final science section, we talk a little more about cDNA cloning and gibberish-containing genes.

At any rate, let's suppose that the genetic engineer—by hook or by crook—has the gene he or she wants cloned in *E. coli,* has identified the clone colony containing the gene, and has succeeded in getting some level of protein expression from the gene. If engineers are very lucky, protein is being pumped out at a high level. Most likely, they don't get that lucky and have a great deal of work cut out to achieve a good yield. Also, *E. coli* may not be the host organism the genetic engineer ultimately wants to use for commercial production. Once most engineering problems have been solved, they may want to develop *B. subtilis,* yeast, or other microorganisms for large-scale commercial fermentation.

To get a higher protein yield, the genetic engineer has several options. First, any way that increases mRNA production will increase protein. To accomplish this, the engineer might increase the number of copies of the plasmid containing the gene and control sequences. Plasmids can exist in cells in multiple copies. The number of copies depends on one of the plasmid's DNA sequences, a signal that tells the host cell to make another plasmid copy. This signal region is called the origin of DNA replication. Remember that strong promoters help begin transcription of mRNA. Analogously, if the origin of DNA replication is strong, the plasmid will reproduce more often, and more copies of it will be available. Recently, genetic engineers have isolated strong origins of replication from plasmids, identifiable because the plasmids containing them are always found in many copies in the host cells. These origins can be isolated and spliced into a quite different plasmid that contains the desired gene. More plasmid copies, more mRNA, more protein.

Another method of increasing protein yield is to find strong promoters, as mentioned earlier. A relatively simple way to do it: locate the proteins produced naturally in the largest quantities in *E. coli,* identify their coding genes, then search "upstream" from those genes for what ought to be strong promoters. By studying promoters, engineers can learn what makes a promoter strong or weak. Then they can splice parts of various promoters together and arrive at a hybrid promoter stronger than those found in nature. A few such artificially strong promoters already have been constructed in *E. coli* to increase mRNA synthesis for those messengers that lead to important proteins.

Still a third method of increasing protein production is to build better ribosome binding sites on the mRNA. The sequence of the basic ribosome binding site for *E. coli* is known already, but there is increasing evidence that the base sequences surrounding that site not only affect binding but may also affect the *rate* at which translation begins.

Combining all these tools and tricks of the genetic engineer's trade, production of a desired protein can now as a general rule be increased to 20 percent of the total cellular protein output. In some cases, genetic engineers have gotten yields as high as 40 or 50 percent, and one firm claims it has gotten a yield of 70 percent for one protein. At 50 percent, the yield is 500 times that of the normal cell. At these higher yields, the *E. coli* colony may be killed by its own overproduction, but economically that may be worth it for such a high quantity of protein.

There is a catch. Even when all the promoters and other aids are perfectly placed, some proteins are not expressed in quantity. Sometimes they are degraded (destroyed) as they are made by enzymes called proteases, which are present in all cells. For some reason, *E. coli*'s proteases degrade some genetically engineered protein products before they can be isolated from the cell, but not others. As a rule, it seems that smaller proteins such as hormones tend to be degraded faster. Consequently, experimenters have been trying to develop *E. coli* strains that are protease deficient, so that degradation of the product will at least be reduced.

Other Microfactories

We conclude our discussion of the fundamental science of genetic engineering with a look at two other interesting organisms that may become rDNA factories, *B. subtilis* and yeast. They are interesting primarily because of valuable enzymes and pharmaceuticals they can help produce.

Consider the previously mentioned industrial enzyme α-amylase, which turns cornstarch into sugar that can be fermented into alcohol. Genefacturing the enzyme in *E. coli* requires destroying the colony to get the product. But in *B. subtilis* the α-amylase will be secreted out of the cell into the surrounding medium. We would then spin the cells down in a centrifuge, concentrating the cell solids into the bottom of the centrifuge tube, and pour off the liquid with the nearly pure enzyme, saving a lot on purification costs.

The general principles of genetic engineering remain the same in these organisms, although details differ. For example, the promoter DNA sequences from *B. subtilis* will work in *E. coli,* but those from *E. coli* will not work in *B. subtilis.* That is especially unfortunate because it means all the strong promoters collected in recent years for *E. coli* cannot be used in *B. subtilis.* Nevertheless, scientists have begun to isolate strong promoters from *B. subtilis* and to learn why promoters differ between the two organisms. Ribosome binding sites also differ. While it took many years to develop sophisticated genetic engineering technology in *E. coli,* it should only take a few years to develop parallel technology for *B. subtilis.*

Even more importantly, we will have to learn from scratch how *B. subtilis* secretes its proteins through the cell membrane into the medium. Apparently, *B. subtilis* secretes certain proteins because they carry amino acid "signal sequences" that are compatible with the molecules of its membrane, allowing the protein to enter the membrane and pass through it to the outside medium. Just before exiting, these signal sequences are usually sliced off, so scientists do not know what many of them are. Once known, the signals could be tacked onto genes for almost any protein so that they too could be secreted into the medium uncontaminated by the cell's other proteins. Although *B. subtilis* is still seen as a potential organism for industry, it has been slower to yield its secrets to molecular biologists than have other cells, including cells of higher organisms.

Yeast, of course, is an alcohol producer. In addition to the value of alcohol already discussed, genetically engineered yeasts may be used in the alcoholic beverage industry to make so-called "light beer." Regular beer contains so many calories partly because the starch from the grains fermented into beer cannot be metabolized by yeast, which lacks the proper enzyme. Unfortunately, we don't lack the enzyme, so when we drink beer, we metabolize the starch into high-calorie sugars. Result: the infamous beer-belly.

Light beer now is made by dumping into the fermenting brew a batch of enzyme that metabolizes the starch. The 90 to 100 calories of light beer versus the 150 calories in regular beer represent degraded starch. One genetic engineering company, BioTechnica International, has now spliced the gene for this enzyme into brewer's yeast, meaning that the yeast itself digests the starch, with no additives needed. Now the yeast, rather than you, will get fat. Whether the resulting beer's taste is improved remains to be seen; if it only remains the same, a company using such yeast would at least be able to say its beer had no ingredients other than water, yeast, malt, and hops.

So far, we've only talked about yeast's production of alcohol, but it could also be used eventually to make products that cannot be made in *E. coli* or *B. subtilis*. Certain proteins of higher animals *and* yeast undergo modifications after translation on the ribosome. For example, some proteins have sugar residues tacked onto them in various spots. Bacteria never make these so-called "post-transcriptional" modifications and don't have the apparatus to do so. That means any animal protein that requires these alterations cannot be made in bacteria. Early experiments with yeast have not been very successful, because it turns out that the yeast tacks on slightly different sugars than animal cells do, so the resulting proteins frequently remain nonfunctional. If these problems are not overcome, yeast still remains an attractive organism to contemplate for industrial genetic engineering because its fermentation is so well understood.

If so many problems loom in trying to engineer animal genes into bacteria, why not forget bacteria and go right for the animal cells as host? Why not place interferon genes within animal cells in front of strong promoters to make interferon plentiful in its native habitat? Major advances are making this possible, but let's consider the drawbacks.

Cells of higher animals grow in a very protected environment, bathed in nutrient-rich blood and maintained most of the time at well-regulated temperatures. It is possible to alter them so they will grow in a nutrient medium in a laboratory dish, but they are very finicky. Their growth medium must be rich in vitamins and nutrients, and that means it is costly. *E. coli* will grow well on sugar water with a few additives. By contrast, the standard medium for growing animal cells is the serum of fetal calves.

Growth rate is even more serious a problem. *E. coli* and *B. subtilis* double every twenty minutes, but an animal cell doubles once every twenty-four hours. That means in ten to twenty hours you can grow up a fermenter-full of billions of *E. coli* cells that might yield hundreds of pounds of protein. The animal cells will have doubled just once in that time, and it could take weeks to get a significant yield of protein product.

The long-range potential for engineering directly in animal cells is therapeutic. Lack of certain proteins or mutations of existing proteins causes many genetic diseases, including sickle-cell anemia, Tay-Sachs disease, and beta-thalassemia. If you could engineer into the patient the gene coding for the protein, you might cure the disease. Certainly, there are scary aspects to this kind of genetic engineering: you might alter humans for other than therapeutic purposes. Recent experiments in creating transgenic animals certainly have brought us closer to genetic engineering in humans.

Such experiments indicate the tremendous progress that has been made in mammalian cell genetic engineering in the past few years. Many molecular biologists believe that within a few years many drugs and specialty chemicals will be produced in animal cells or—an even more dramatic advance—in living animals.

Plant genetic engineering, though still in its infancy, has advanced beyond the expectations of most genetic engineers just within the past few years, and the social and economic potential for plant genetic engineering is staggering. As it develops, we are also going to have to consider the hazards suggested by some scientists, such as the possibility of altering nature's balance.

5 BULL, BEAR, AND CLONE

Venturers

Life generally proceeds by small changes rather than by cataclysm, whether the changes are planned or accidental. An alteration in the way things usually occur will produce an effect small or large, and life will either follow the new course as a superior/easier one or ignore what then becomes aberration. Form follows function, for the "best" form among a large number competing for acceptance is the one that *works* best.

That might pass for a business view of evolution or an evolutionary view of business: orderly, nonradical, a survival-of-the-fittest scheme in which corporate form follows product demand, in which laboratory discovery takes root in pilot project, then blossoms into a "big boomer" in the test markets to yield an industrial giant—or is ignored and never comes to be at all.

It is an especially apt description of American business strategy, which compared with others in the world—notably Japan's—is considered short-range. American businessmen seek fast returns on investment, and that was a major consideration five years ago as venture capitalists and market analysts were forecasting a gloomy future for the new genetic engineering industry when its startup firms failed to produce a fast dollar. They have, indeed, not paid back most of their investors. But the crash never came. Even the most volatile public market has been patient beyond any other time in the memory of those who make a business of predicting other people's behavior.

Now consider genetic engineering, the largest collaboration ever between business and biology and one that in less than a decade already rivals the similar fusions that created the chemical and computer industries. Unlike these others, biotechnology was born in direct violation of virtually every evolutionary notion. Evolution implies gradation. Genetic engineering was born as an industry fully formed on October 14, 1980, the day Genentech went public and set off a wild craze for biotech stocks that grew, ebbed, and now has surged again. From fewer than a dozen companies in existence on that date, the industry has grown to three hundred companies set afloat on a combined investment of $3.5 billion.

Perhaps most surprising of all, although returns on this investment have been slower than predicted in a country whose business strategies are notoriously short-range, the transfer of technology to the marketplace has been faster. The products of genetic engineering are beginning to reach the market, and they range from pure replacement drugs, such as human insulin, to trace body chemicals, such as human growth hormone and interferon, that are being tested and sold as new drugs. Nevertheless, it's safe to say that the investors in genetic engineering are still riding on faith. These companies are ultimately depending on orderly earnings growth, and on a super "score" that can only come to some—without which prospects for independent survival are certainly diminished.

Is this any way to do business? Maybe so, because this industry is a creature of venture capital, a uniquely American institution that has been quietly building since the end of World War II. Venture capital nurtured the transistor and electronic data processing from birth through entry into the small business and home. But it truly can be said that venture capital never had created an industry before. Of the $3.5 billion estimated invested in biotechnology, venture capitalists account for about $2.25 billion.

So who are these venture capitalists in this age of entrepreneurs? Morton Collins, who provided the investment figures, offers this apt definition: Christopher Columbus was the prototype entrepreneur. He had an idea which he believed would pay handsome returns if put into action. Queen Isabella was his venture capitalist, bankrolling him in the hope that she would get back far more than she invested. The analogy is even more apt, Collins points out, considering that Isabella's investment paid off far more handsomely than anyone could have dreamed—and Columbus's idea didn't turn out the way he'd planned at all, in his case because it was the Old World rather than a new one he'd envisioned and thought he'd found.

The notion of venture or risk capital runs through most free enterprise economies. It refers to the money you can afford to lose, money you can therefore bet on something with long odds, but which would yield a really big score if the project connects.

Thomas Perkins takes some exception to that common definition, but in so doing he puts a finer point on it. The chairman of the board of Genentech and the first venture capitalist to put money in the field, Perkins says: "Venture capital does not go after high risks. Venture capital seeks high returns. High risks just tend to go along."

The sources of venture capital are varied, ranging from the wealth of America's most powerful families to some corporate funds that are plowed back into the field—instead of into taxes, for example.

Perkins himself is a perfect illustration of how things get started. A graduate in electronics from M.I.T. with a Harvard MBA, Perkins rose in the corporate ranks of the budding electronic data processing industry. He started Hewlett-Packard's computer division, then became its general manager. As part of his job, he oversaw the creation of Optics, Inc., a firm in the even-newer field of laser technology. There he got a bright idea and brought it back to H-P with him. He conceived a laser which would lack the fine-tuning refinements of laboratory models, but which might be used for "dirty" jobs—inspecting pipes on construction sites, for example—jobs that would involve wear and tear, but would not require the precision of the scientist's tool.

He remembers asking H-P "if I could carry on my little hobby on the side, of developing this idea I had, and they said I could. That was pretty unusual for a company to grant someone at my management level."

Looking back, Perkins says he was at that moment a venture capitalist, but didn't know it. "What I did was to assemble the people and the money to bring this idea to reality," Perkins said. The result was what he calls the lasertron, the world's first "light-bulb" laser that simply switches on and off without tuning. "We did very nicely with that," he says modestly. He is a senior partner in the San Francisco venture capital firm of Kleiner, Perkins, Caufield and Byers, whose spacious offices overlook the Bay and Marin County from the upper reaches of an Embarcadero tower.

A soft-spoken man, he says with more than a touch of pride, "If we hadn't backed Genentech, I'm not sure the industry would have formed the way it did. I'm not sure where it would be now."

Indeed, what might be a boastful statement coming from anyone else is an understatement from Perkins. He and his firm have been virtually the fountainhead of genetic engineering financing, from the birth of Genentech, to its public offering, to the creation of Hybritech, whose purchase by Eli Lilly in 1985 would put a high valuation on an industry that still exists to a large degree on paper.

Genentech is not the oldest of the genefacture firms, but it was the first founded purely to do recombinant DNA: the central, most elegant, and in the long run, most promising art of biotechnology, and what we have been referring to as genetic engineering.

Venturing, basic version: man with capital meets scientist with idea. Realizing that each has something the other can use, they put the one's cash and the other's superbugs together and hope for the big one: the patentable bug that spits a miracle drug all day and multiplies like crazy—or some variation on this. Herbert Boyer met Stanley Cohen and the idea of gene-splicing as a new technology was born, but there was still no industry. It was when Boyer met Robert Swanson of Perkins' firm that skyrockets went off. Swanson and Boyer each put up $100 of their own money to incorporate and Swanson hit Perkins for $100,000 in startup money. "He was hot as blazes on genetic engineering," Perkins recalls. "He was going to make a career out of this, either with us or without us."

Risk? "Very high. I figured better than 50-50 we'd lose it. But it's rare when the odds on a new technology are better than 50 percent."

Second thoughts? "Not at all. If it worked, the rewards would be obvious."

Boyer and Swanson had a strategy for quickly developing products such as insulin, human growth hormone, and other drugs that could be sold to existing major drug companies with ready markets, so it was no accident that the company teamed with Eli Lilly to develop human insulin. This short-term revenue-producing plan was to be matched with a longer term strategy of becoming a fully integrated pharmaceutical company—researching, developing, marketing. Finally, the two saw Genentech farther down the road as a challenger in the capital-intensive field of manufacturing industrial chemicals not now made through biotechnology, a more revolutionary use of genetic engineering they felt would be unlikely to produce immediate revenues.

Swanson remains Genentech's president. The company is consistently rated at the top of the industry by market analysts and held up as a model of how a genetic engineering firm ought to operate. Genentech marketed the first product of genetic engineering, human insulin, in 1982; sold several more products since, and now has two candidates for the first "boomer" of the industry: human growth hormone, now in clinical trials, and Tissue-specific Plasminogen Activator, reportedly far better at eliminating blood clots after heart attacks than other drugs. The first company to go public, its stock is now valued at $81 a share after having split twice. And this before it or any other genetic engineering firm ever became what a businessman would call profitable.

At the moment Perkins was giving Genentech its $100,000 launch, on the other side of the country Robert Johnston was getting excited about the field. A former investment banker who had been putting up money for and running various medical equipment firms in a business run from his Princeton home, Johnston was in regular contact with biochemists, molecular biologists, and medical researchers.

"I guess what I couldn't believe was how excited *they* were by the technology," Johnston said. "Scientists, you know, are a pretty conservative lot, especially when it comes to guessing how practical the application of their discoveries will be. To some, it's almost a dirty word, 'practical.' But all of a sudden, everywhere I went, *scientists* were enthusing over genetic engineering and what it could do."

That period, the late 1970s, also marked the height of public controversy over genetic engineering and its feared risks. Many of the questions raised about recombinant DNA were valid. But, not surprisingly, a heat storm erupted around the legitimate questions simply because threats are perceived in *any* new technology. In that adversity, entrepreneur that he is, Johnston saw opportunity.

"The major pharmaceutical houses were holding off [putting the technology in place] because of the controversy," he recalled.

"Pfizer, one of the country's biggest houses and one of the most experienced at fermentation technology, had been picketed in Connecticut, and they had stopped in their tracks. It seemed to me a good chance for smaller companies to step in, develop the technology, and sell some useful products competitively."

Johnston's perception was prophetic. Nearly a decade later, many others in the field as well as in the investment community believe that the genetic engineering industry grew up independently, with ties to the pharmaceutical and chemical giants rather than as part of them, because of that relatively brief period of hesitation in the late 1970s, and because early on the smaller firms could attract top "bench" scientists who are still relatively rare as employees of giant firms.

Unlike Genentech, Johnston planned to develop genetic engineering techniques for producing amino acids, for some of which there was already a large, lucrative market. These protein building blocks, as we pointed out, are useful nutritional supplements, especially in animal feeds. They were already being produced by microbial fermentation, so to enter the market one needed only to build a better microbe. Johnston had no microbe engineer, but what is an entrepreneur if not enterprising?

He put an ad in *Science* magazine, the official organ of the American Association for the Advancement of Science. "Entrepreneur seeks scientist to be president of genetic engineering company," the ad read, then went on to detail plans and requirements. Johnston got about half a dozen replies, but "most of the respondents were unqualified."

Only one applicant completely filled Johnston's bill of particulars—Dr. Leslie Glick, who not only became Genex's president but later founded a trade organization serving biotechnology companies. Johnston says, "You usually think of mad scientists and conservative businessmen. We're kind of a reversal—mad businessman and conservative scientist." It was Johnston who had to convince a doubting Glick of the enormous promise in the field.

The nearness of Genex, in Rockville, Maryland, to Washington aided the company in becoming a major collaborator in the federal Office of Technology Assessment study on the future of applied genetics. Stock analyst Scott King, who helped draft the report, says that the collaboration was too cozy, given the company's interests in the field, but a Genex board member counters that the firm was the only one that had the technical resources to provide the data the government needed.

Johnston recalls hard early days, a year in which the company existed virtually only on paper as he and Glick tried to scrape up funds. But he sees that as a boon to the company's corporate toughness. "It's better to grow up lean, appreciating the value of a dollar," Johnston says. "A lot of computer companies in the sixties went down because they got way too much money when it was plentiful, developed bad habits, and could not survive when the crunch hit."

That crunch, totally unknown to or forgotten by most people, is a major landmark on the rough economic terrain of high technology, and is frequently referred to by those now assessing genetic engineering. Like biotechnology, electronic data processing (EDP)—the computer industry—was born on waves of futurist fanfares that brought millions of "easy" investment dollars before the bubble burst in the early 1970s. A few of the best EDP firms survived into this "golden age" and are now making a fortune, but hundreds of others did not. In the early 1980s, many if not most stock analysts foresaw a similar fate for the vast majority of new genetic engineering firms. The firms have survived the first "bust" in mid-1983, and most of those who watch the field believe that we are entering a period of consolidation through acquisition, when larger, stronger firms will take over the small and weak. But there is disagreement on whether the terms will make this period a happy or unhappy time for the acquiree's investors.

Genex began and ran for several years as a privately financed firm, then went public in 1982. There is a vast difference between public and private financing, and each has its pros and cons. Johnston pointed out that public money "is the cheapest money available." The corporation pays far less back in earnings on the public investment dollar than it must give away in equity in return for venture capital or private financing arranged through investment bankers. Investment bankers broker stock offerings to private investors, but they generally do not take part in managing the company. Venture capitalists do.

But public money has drawbacks. Morton Collins notes, "Public money is impatient, demanding, and volatile. When the public likes you, the money flows. But when the public and the Wall

Street analysts decide it's time you showed a profit, the money stops quicker than any other kind."

Whether public or private, most genetic engineering firms have venture capital somewhere in their financing history, and that in itself is new. Collins, former president of the National Venture Capital Association, a 200-member trade organization and lobby, said that the concept only solidified in the last fifteen years, and that at its inception it was not at all specialized with respect to areas of investment. In those early days, Collins said, he foresaw a need for seed money in the booming data processing industry—and if there is a phrase that describes venture capital best, it is "private seed money." Collins founded Data Science Ventures. At about the same time, two friends were trying to start a similar "seed money" firm to back companies specializing in medical systems. The time was not yet ripe for Peter Farley and Ronald Cape, Collins relates, but they went on to develop Cetus.

But venture financing not only brings money, it brings management controls. Entrepreneurs such as Johnston typically put together business plans, then look to venture capital for financing. A given firm might offer $2 million in return for half the company, *and* provide a certain number of directors for the firm, possibly the board chairman. By these actions, the venture group has put a value on the company based on its experience of corporate worth, and it provides expertise in running the business as well as the infusion of money that starts it off.

Collins says, "We've done more than sixty start-ups in fifteen years. When we come in, our executive expertise is worth more than the money we're bringing, and we try to impress that on those we deal with."

In their turn, what venture groups think is of highest value in the companies they back is personnel—particularly if the company represents high technology, the area in which venture groups have done most of their financing. Johnston says, "Most venture people will tell you management is 80 percent of the worth of your company. That means business executives and your top scientists."

Venture groups say they also bring a measure of stability to public offerings, in the interest of protecting themselves. Gerald R. Lodge, whose InnoVen had its own key role in the industry's creation, pointed out that the Securities and Exchange Commission has relied on adequate disclosure of corporate information to protect the stock-buying public. If venture groups were to sponsor offerings that turned up consistent losers, tougher regulations would be inevitable, "and that's the last thing we're interested in."

Lodge is chairman of InnoVen, perhaps the only venture group in which corporate investors play a working role. In 1974, as the industry's day was just beginning, Lodge, a partner in what was then a small venture management group, was contacted by the Monsanto Corporation and Emerson Electric. He was asked to "meld our capital management with their technical expertise" in a series of investments of corporate funds and some private investors' money. Lodge agreed, and a year later found himself at a Monsanto tutorial that focused on genetic engineering.

"The gist was that this was the field to watch, but no rush about it. The corporation didn't think we'd find anything to invest in for several years and believed the technology would not yield sales and profits for another ten years."

Lodge, then on the venture association board, ran into Perkins at a meeting only a short time later and heard about Genentech. "I want in," Lodge said, and InnoVen made two large investments in the company as it was starting up.

Monsanto executives were delighted with the investments for two reasons—they saw a "window" on the future, and they foresaw a necessary hard learning experience on the horizon. "They said they were glad we got in so early, but basically: 'You're going to lose your ass. The start-up funds will be liquidated before the company makes a nickel.' "

That forecast, like so many others in this field, proved greatly mistaken. By 1981 Genentech had produced revenues of an estimated $11 million on contracts with Ely Lilly for insulin, Sweden's KabiGen for human growth hormone, and Hoffman La Roche for interferon. The firm reported earnings per share—$0.02—for the first quarter of 1983, and its stock was selling at $42.25 in early May. Genentech's prospects with its new Tissue Plasminogen Activator could make these sums look puny, as we'll see.

But Lodge remembers a "uniformly conservative view" by most scientists as the decade began. A topflight medical researcher gave Monsanto a list of eighteen technical hurdles that had to be overcome before recombinant DNA would have an impact on industry. He projected a five- to fifteen-year time lag for every product, based on when the various hurdles would be cleared. "All eighteen challenges were met within two years," Lodge said, an indication that the products are much closer at hand than originally forecast.

In the summer of 1977, Johnston and Glick approached InnoVen for funding. The "storm and heat" over hazards were then at their peak and Lodge initially was hesitant, but the firm eventually put up "several hundred thousand dollars" for about one-third of Genex. Lodge made it clear that he was interested in Genex's ability to apply genetic engineering to industrial rather than pharmaceutical products. In effect, this meant Genex was to prove its skills at reducing costs of making existing chemicals rather than concentrate on new inventions, such as wonder drugs. This illustrates the management side of venture financing; venture capitalists generally either help new companies to get up and running, or they infuse cash into existing firms with exciting ideas for which they need funding. Within a short period—typically three to five years—the venture capitalist takes leave, and the fledgling must stay airborne on its own. That means producing a revenue stream and then, perhaps, going public. Or, in the strange and unique case of genetic engineering, doing it the other way around.

Building from Scratch

Mort Collins lives just down a country road from Bob Johnston—so close, Johnston says, they can wave to one another when the foliage is off the trees. Nevertheless, just a few years ago they did not see eye to eye at all, at least on the future of the genetic engineering industry. Collins was among the most pessimistic concerning the corporate survival of the "genies," and his reasoning is instructive about the nature of business building in America.

We mentioned, for example, that venture capital is an American institution. On the one hand, European businessmen and scientists frequently lament that they have no such funding available to them on any regular basis, and many believe that is the reason that most European genetic engineering development lags behind the United States'—though perhaps not for long, because governments in some countries have stepped in to provide upfront financing. The Federal Republic of Germany, for example, in 1985 decided to infuse 1 billion marks—about $300 million—in West German biotechnology. This is more like the Japanese model; for, on the other hand, the Japanese government participates intensively in goal-setting for industry and science, and in contrast with the United States, the Japanese are noted as long-range business planners.

What does this mean to someone starting a new industry? To Collins, the watchword is "rational": rational behavior, rational planning, rational expectations, and above all, rational and quick performance. Collins was a survivor of the collapse of EDP, remember, and he recalled that in the late 1960s, "All you had to do was say you were in computers and the money flowed. This was not rational investment, this was not rational economics. By the early 1970s, you couldn't sell a share of computer stock. Companies needed money for future activities—it wasn't there. Most of them had developed very bad habits, they weren't used to making a little go a long way. They're gone."

The immediate cause of the collapse, Collins believes, was a failure to produce a rapid flow of regular earnings. "Somebody gets the idea you're not sound. The idea spreads—to good companies, bad companies, any company that can't show that stream of earnings—*and none of them can.* Soon a portfolio manager looks like an idiot if he hasn't dumped the stock—and *that* is the opposite of being hot."

Collins foresaw even then that some of biotechnology's majors, such as Genentech and Cetus, had enough cash to survive a falling out with the public until that steady stream of cash would come in. Nevertheless, he said in 1981, "I predict that soon you won't get venture capital to invest in this business that venture capital started."

Nor was Collins alone. Scott King, then the biotech stock watcher at F. Eberstadt, and Joyce Albers, who covered biotechnology at The First Boston Company, both saw a shake-out on the horizon that would equal if not surpass that suffered by EDP. Both were trained in biology: Albers has a degree in microbiology from Rutgers, King a master's in enzyme chemistry from Harvard. Even after the "over-heated, circuslike atmosphere" in the summer of 1981 had settled down and King's worst fears ended, he and others believed that a major tremor was coming. Scientists in the field expressed the same concerns.

"Shake-out" was among the most commonly heard terms in company offices, each player trying to guess when that period would come when the easy money would disappear and the demands to produce would overtake eagerness to invest in the newest game in town. Company managers in interviews frequently spoke of a given contract or project as good enough to carry them through the shake-out—and they worried aloud about poor operations that had sprung up that were coloring the public perception of the entire industry.

The *Wall Street Journal*[1] ran a story about one firm that consisted of two well-qualified scientists and a board of directors that included a disbarred lawyer and another man with suspected organized crime ties. As a result of its story and other publicity, the *Journal* reported two weeks later, the scientists bought out their suspect partners.

Venturist Lodge said bluntly that he feared the credibility of the nascent industry was being threatened by "some real flimflam proposals I've seen, with some major investment houses backing them."

But for once, the worst apprehensions did not materialize. By the end of 1981 there were about 150 companies carrying out genetic engineering in one form or another, and that year they got public financing totaling nearly $140 million. In the first *quarter* of 1986 there were nearly 300 such companies (though the lines have blurred a bit between genetic engineering and other forms of biotechnology), and in that quarter the public companies got $324.6 million, all but $57 million by public sale of stock. What happened to the shake-out? The tremor occurred, all right, and it occurred precisely at the moment Collins had forecast; it simply didn't have the effects predicted. Reason: No one had anticipated the enormous amount of money venture capitalists, giant corporations, and the stock-buying public were willing to put up on faith in a new industry. When the loss of confidence occurred, in mid-1983, virtually all the companies had enough money in the bank to survive.

Scott King now says he underestimated investor interest, but that product development has also been more rapid than he expected. But on at least one score he has not changed his view. In 1981, as an example of the vagaries of market forecasting, he talked about the potential impact that a cure for diabetes would have on the human insulin market. A couple of years later he suggested that a cure for diabetes would be found within ten years. He has now left security analysis to start up a company trying to cure diabetes. Its goal is to transplant the pancreatic cells that produce insulin into diabetics, using new methods of treating the cells to prevent rejection.

Meanwhile, when Collins is asked about his own dark forecast, he answers with characteristic candor: "First, I was wrong." But what comes second and subsequently is far more interesting: "I still hold by this principle: The marketplace will believe whatever you say, but then you must perform or it punishes you very severely and without forgiveness." That's what caused a very sharp drop in financing to genetic engineering in 1983, when an Amgen public offering failed to sell out and a subsequent major offering by another firm was held up as its offering price was cut; then that offering, too, failed to sell out. Cetus's initial public offering also had not sold out and its price had fallen, two years earlier, but Cetus had raised $125 million, enough to weather any market storm. Collins marks those 1983 events as the "breaking" of the public market. But not only did all the companies survive, they rebounded into public favor, still without huge revenue producers.

Most recently, a Shearson Lehman Brothers report on biotechnology gave Amgen top marks among genetic engineering firms.

Says Collins: "These companies have not performed—that is, created a large revenue stream—in a far longer time period than the market waits, yet they have not been punished; they have even come back into vogue. If you look at only the value in the public market, it's astronomical compared to what you could rationally imagine. Companies with a lot of cash are much more numerous than those without cash."

New biotechnology companies raised nearly $3.6 billion between 1980 and 1985, Collins estimates. He breaks the financing down this way: venture capital, $2.24 billion; public equity, $769.1 million; corporate equity, $198.6 million; and R&D partnerships, which we'll talk about below, $355.1 million. The public equity sales for the first quarter of 1986, mentioned earlier, bring the total of publicly held equity to over $1 billion.

Collins says that in trying to gauge the industry, he initially underestimated the public's willingness to try something new. But additionally, a new set of alliances came about that helped the start-ups stay afloat.

Congress created a program known as the Small Business Innovation Research (SBIR) program, and among other features it requires federal agencies to give a certain percentage of research and development grants to small businesses, of the size of the genetic engineering start-ups and other small high-tech firms. The program is highly regarded, in part because proposals are scrutinized by the same sort of peer-review committees that review basic research grant applications throughout the federal system—except that industry representatives are included on SBIR committees as peers along with scientists.

A new type of company–investor alliance referred to by many corporate officers is also contributing in a major way to success: the research and development partnership. Basically, it permits a company to carry out a project it could not otherwise afford with investor money specifically linked to the project—that is, no company equity is involved, unless a small amount is added as what Collins calls a "sweetener." The key to this funding scheme, however, is that the partners can write off their investments as tax losses if the project does not prove fruitful. That is expected to change under the new tax law and may mark a practical end to these partnerships.

Finally, Collins points to the unusual nature of giant pharmaceutical houses as contributing to the success of startup companies. Unlike, say, automobile companies, they do not compete with each other across the board, but the nature of their business may indicate something of the future of biotechnology. Each drug company has one or a few product areas that account for the bulk of its business. "You may be looking at a $4 billion company," Collins says, "but $3 billion is in antibiotics. The bulk of their R&D spending, which is enormous, goes into the area of their specialization, not toward broadening their base."

Few saw what was happening as genetic engineering came onto the horizon—Monsanto, Johnson & Johnson, and Eli Lilly being among the few—but even for those who saw what was coming, the startup companies offered a good way to get what Lodge calls "a window on the new technology." If a firm's expertise lay in a different direction from producing interferon, then it could pour money into a promising start-up that already had top scientists on staff and on the board of advisers. The nature of the pharmaceutical business, Collins says, "created unexpected availability of capital for both equity investments and R&D contracts on very attractive terms."

Why does the drug industry behave as it does? Partly because patents play a major role in the business, so once a company has exclusive rights to a drug, it puts all its marketing skills into it. Peter Farley, while still chief executive officer of Cetus, pointed out that most pharmaceutical houses live for most of the twenty-year life of a patent on the income from one or two proprietary antibiotics or other wonder drugs. During the end of that period they begin searching intently for where their next "boomer" will come from. When the proprietary drug goes off-patent and becomes generic, the price drops and the firm expects little if any additional income from it. And probably of equal importance in explaining why drug houses concentrate as they do is the cost of developing a new drug, estimated at $30 million and up, largely because of FDA testing requirements.

To take a contrasting example, computer firms also rely heavily on patents for protection, but they are entirely different from rDNA firms in one vital aspect. As Tom Perkins put it: "There is no FDA in computers." Thus, the history of that industry is replete with tales of products developed in engineers' garages in their spare time, with small startup investments leading to fabulous riches. Genetic engineering is capital-intensive in almost every aspect in which its products are intended for human consumption or environmental release—and that means many, though not all of its applications.

Perkins pointed out that Ph.D. scientists are this industry's "limited resource," and that fact probably will keep firms into pharmaceuticals and other high-value-added products, rather than bulk chemicals in the foreseeable future. Bulk chemicals could be made with less energy and toxic waste than using current methods, but, "You want the Ph.D.s working on products selling for thousands of dollars an ounce rather than a couple of dollars a ton," he said. "The major reason that bulk chemical production via genetic engineering is probably in the distant future is an economic one"—although there are many specialty chemicals that fall between such extremes and would be of value for a company to develop.

What is the future of the start-ups? Collins and Perkins both see a long succession of acquisitions coming, although, interestingly, Collins is now the more optimistic. "I think in most cases, these acquisitions will be on very favorable terms, terms which will make the investors very happy to be acquired." Perkins sees that in some cases, but for many weak firms he believes larger companies will be buying up their research "at less than a dollar on the dollar."

Nevertheless, both feel venture capital will continue to play a role in genetic engineering, though certainly not the role it did in the past, since both private and corporate money are increasing. Collins, whose firm has put venture money into four biotechnology start-ups, noted, "Many products are nearing market, and that should mean the start of major revenues for the start-ups as well as their partners. Humulin was the first to market, for Genentech and Eli Lilly, but with HGH coming out, Genentech is nearest to having a real, new product on the market. TPA is coming, and a myriad are not far behind—interleukin-2, the interferons. Some are proprietary and are going to be very important revenue sources."

Collins believes the day is near when "I'll go into the doctor's office once a month and leave behind a few drops of blood, and that'll be all it takes to tell if I'm in perfect health or not." And Collins is no enthusiast, coming new into high technology. He holds masters and Ph.D. degrees in chemical engineering, a second master's in mathematics, and spent seven years teaching at Princeton, falling into electronic data processing almost by accident when he became involved with the university's computer center.

And even in the period when Collins' view of the near-future of genetic engineering was at its most "bearish," he remained strikingly optimistic about its eventual impacts. Standing in his Princeton yard on a summer evening in 1981, predicting an apocalyptic shake-out, he enthused without skipping a beat: "Genetic engineering has the capacity to bring about a more profound change in our society than we have ever seen before, one that enters not only industry, but strikes at the moral and religious aspects of our lives as well. We will be toying with immortality some day, no doubt of it. I don't know if we can deal with it, but it will be there. We're talking about a fundamental understanding of the life process. We're talking about a dramatic potential for good and evil."

More than from nuclear physics or computer science? "I think so, yes."

In the summer of 1986, he added, "Its potential is more tremendous to me now that I know more about it."

Two Companies

> There's something strange going on here. The scientists are making money and the companies are publishing.
>
> —Wife of a scientist

Heading up toward San Francisco on Route 101 with the sky full of clouds but the autumn air remarkably clear, you turn off the flat peninsula into South San Francisco's heavily industrial Point San Bruno, along the edge of the Bay. There, in a low-slung sprawling modern building sits Genentech, wearing its corporate name tag modestly low along the shrubbery. Within its single building work hundreds of scientists, technicians, executives, secretaries.

Genentech's headquarters have undergone an interesting metamorphosis in the past few years, reflective of what has occurred to the company and the industry in which it is a pioneer. At first, behind the entrance reception area, closed off by display glass, there was a corporate badge of identity: a gleaming 750-gallon fermentation tank, working model. If this was meant to show up front what Genentech was all about, it also emphasized the smallness, the compactness, of the operation. Fermenters double that size are commonplace, and industrial fermenters have been built for genetically engineered products in the 25,000-gallon range. But perhaps more significantly, Genentech's fermentation is now done in a separate manufacturing plant; that headquarters space was remodeled into laboratories.

In the brief history of genetic engineering, no date has more importance than October 14, 1980, the day Genentech went public. That marked a turning point in recombinant DNA as industry, a milestone to compare with Cohen and Boyer's collaboration in the science, and one that illustrates a major feature of public money: it carries with it publicity, which brings new industries out of the purview of a few to the scrutiny, interest, conversation of many—for better or worse.

Genentech raised $35 million in a few hours in return for only 13 percent of its equity as its offering sold out, then was traded wildly for the remainder of the day. Offered at $35 a share, the stock quickly rose to $89 before closing just below that. At the time it was the largest single public stock offering in history. Genentech was the darling of Wall Street, the subject of countless newspaper and magazine cover stories and of a *Time* cover[2] less than six months later.

That boom represented the sprouting of a technology from pure science, and never before had the push from the laboratory occurred in such record time. In no other case has an entire cutting-edge science been translated into technology in a few years. Not surprisingly, such an eruption can bring financial headaches and even ruin for some who misunderstand what is happening, be they investors or scientists. But the speed of the transfer is also bringing the promise of genetic engineering to fruition faster than anyone imagined just a few years ago. That occurred in part because a few business executives and scientists virtually reversed roles, the scientists learning the language of the business plan, earnings per share, profit–loss, and the business executives putting money into pure research—always considered a very risky gamble.

Of the early days, Perkins says, "When we founded Genentech, we were backing pure research. At that time, that's all it was. I'm not saying genetic engineering wouldn't have started without us. It would have, of course. But I don't think anyone else would have given it the push we did." As we'll see, that push didn't go quite the way Perkins had planned.

If after leaving Genentech's headquarters, you continue driving up U.S. 101 over the hills and past the rooftops of San Francisco to the Oakland Bay Bridge toward Berkeley, you find in neighboring Emeryville Cetus's corporate offices, and a whole different manifestation of rapid corporate development. A giant hulk that had been Shell Oil's research and development complex has been turned into a $55-million headquarters for Cetus. Nearby a manufacturing plant has been built from the ground up. When we visited a few years ago, we had to speak over the din of construction machinery in what could have been a movie set depicting industrial activity. Cetus was then expanding phenomenally on the crest of *its* public stock offering, which came six months after Genentech's. Cetus's brought in about $107 million to supplant its rival with what is still history's largest stock offering. Cetus suffered several mighty reversals in the years following that expansion, but it emerged in what analysts consider good shape, trimmed down and with its fewer aims better defined. Anticancer drugs are among Cetus's major projects: Interleukin-2, beta interferon, and now tumor necrosis factor.

Cetus and Genentech, the oldest and most established of the publicly held genetic engineering firms, sit like rival siblings on opposite sides of the Bay city. And like siblings grown, they have come to a remarkably similar view of life after often contentious beginnings. Both aim to be fully integrated pharmaceutical companies and are well on their way to being so. Both get very high marks from industry analysts, though no other firm gets the praise of Genentech as the literal model of what a high-tech company ought to be. Both have forced their focus on an area they feel strongest at handling—a tough judgment, because everything is enticing, everything is new, and everything holds out promise for someday bringing enormous returns to the company that midwifes its entrance into the marketplace. Both have resisted the temptation, in fact, by similar strategies, although differently arrived at. They have formed joint ventures, using other companies' money, to explore areas outside their primary targets. That is, they have been able to concentrate their own cash on developing their major strengths while not ignoring other promising possibilities. There are, however, other characteristics that invite the comparison of rival siblings.

Fresh from their public offerings with plenty of cash in the bank, both firms drew large revenues from interest in the early 1980s. Cetus spent a great deal fast, not only in construction in the Bay area, but pushing exploratory fingers into every aspect of biotechnology; Genentech, meanwhile, became famous for carefully culling projects for specific goals, and arriving early at all of its goals. Cetus spread all over the map, making major deals with several large corporations in return for large chunks of its equity, stretching operations all the way to Madison, Wisconsin, where it undertook a pilot ethanol (ethyl alcohol) production plant in partnership with National Distillers and went after a large share of the market in high fructose corn sweetener (HFCS) in partnership with Chevron. Genentech remained quietly rooted in the San Bruno area of South San Francisco and boasts of retaining more control over its equity than any other publicly held start-up.

The completed alcohol project is now on the shelf because of low oil prices; but company officials note that it can be taken from that shelf whenever the price of oil rises sufficiently, as someday it virtually must, because of dwindling world supplies.

The difference in strategies goes back to corporate beginnings, as related on an autumn day in 1981 by Peter Farley, who was then Cetus's president and who had co-founded the company with Ronald Cape in 1972, making it the oldest of the genetic engineering firms. Farley was a medical doctor with an MBA, Cape a chemistry graduate of Princeton who had gone into the pharmaceutical business, then returned to school to get a Ph.D. in biochemistry. From the outset, both Farley and Cape relate, Cetus had in mind exploring the budding technology they saw in biology—and this a year before the first cloning experiments. They had targeted two other areas not directly related to DNA engineering as well: medical instrumentation, since dropped, and screening processes to develop better microbes for antibiotic manufacture. Farley recalled that soon afterward, Nobelist Joshua Lederberg, a science board member, alerted the firm to "very interesting work that a biologist named Stan Cohen was doing over at Stanford."

From that moment, Farley said, genetic engineering was one of Cetus's corporate goals. And now corporate genesis, as Farley related it: At that time, Cetus was partly owned by a pension fund. Then came a stroke of fate, via the U.S. government. The Labor Department decreed that pension funds could not be used to invest in speculative ventures, and Cetus's investors had to sell. The buyer: Kleiner, Perkins, Caufield, and Byers.

In this version of genesis, after a time it became apparent to Robert Swanson, assigned to handle the Cetus account, that he would learn nothing of the inner sanctum, so Kleiner-Perkins sold its Cetus shares and went on to found Genentech. It happened that Kleiner-Perkins sold those shares to a Canadian firm represented by two young men interested in biology and foreign enterprise: Inco. Seeing Cetus's light, Inco went on to put up the major seed money for Biogen.

Farley's story points up two interesting facts about this nascent but growing industry: Just like the science on which it is based, the business was started by a small group of people generally acquainted with one another, and there is no love lost between some of the pioneers.

For his part, Thomas Perkins, ever soft-spoken, says Farley is right that Cetus opened his firm's eyes to the field, and that probably happened because Swanson was assigned to watch it—but Kleiner-Perkins sold its shares in *frustration* that Cetus was sitting on the technology, waiting for it to develop, instead of pushing it toward its birth.

Ronald Cape, a dynamic man whom Lederberg calls the statesman of biology, is now Cetus's CEO and chairman and, as we'll see, an enthusiastic spokesman for the genetic engineering industry and for government support of the basic university research that created the industry. Cape would face difficult years in which Cetus had to lay off more than forty employees and trim back operations, before the company would reemerge as a healthy leader of the industry.

The drive to bring genetic engineering into bloom in the marketplace had its downswings. *Public money is volatile.* No one understands that better than Perkins. In all the hoopla over the Genentech sale, few heard him express disappointment with the way it had gone. After rising to $89, Genentech stock dropped within months to the low thirties. "That tends to make people feel like they got taken," Perkins said. "They seem to feel we reaped some benefit of that price rise." Forgotten by many was the fact that Genentech sold out its million shares at $35—the offering price—and not a penny more. After the shares sold out, demand *among traders* pushed the price unrealistically high; those trading in the stock or buying from traders made and lost all the money as the price rose and then dropped. Many writers looked only at the drop from the unrealistic high in assessing the company, and the industry.

Looking back, Perkins said that deciding the terms of a public offering involves some of business's toughest decisions. Initially, directors and underwriters were going to offer at $20, then they bumped it to $35—a premium price for an over-the-counter issue—because they sensed there *was* a higher demand than initially expected. Perkins still feels the offering price was right, and the price soon after went into the upper thirties and forties—meaning that the "market" quickly priced the company where Perkins had. Now Genentech stock is over $80 a share—its opening day high—and that value must be trebled taking into account that the stock has split twice since opening. What went wrong with the opening then?

Perkins said, "I tend to think we should have offered another

100,000 shares—just another 10 percent. That would have taken a lot of pressure off at the opening; the price wouldn't have gone so crazy." The company actually would have made more money that way, but the point Perkins returns to is: "People wouldn't have felt they'd lost money on us." Further, such volatility in that period indicated to Perkins that people didn't really understand what genetic engineering was about. Now, he says, the market is much more knowledgeable.

When Cetus went public six months later, by most judgments it shot for the moon. The firm offered five million shares at $30 a share. Since Cetus was selling only 35 percent of its equity for that, it meant the firm was valuing itself at $400 million, a figure one long-time biotechnology executive called "completely obscene." But the offering did not sell out, unlike Genentech's. Further, Cetus saw its opening price of $30 fall to $19 a few months later, then to a low of $11 in one year. Since then it recovered to its late 1986 price near $24.

Scott King, then a market analyst, interpreted those reviews simply: "An offering that doesn't sell out, and the stock doesn't maintain something near the opening price? That has to be called a failed offering." But others praised Cetus for the timing of that offering, for it accomplished for Cetus alone what all of them have dreamed of: survival, at least for a few years.

Farley: "Our decision to secure the kind of money we did guarantees—and I mean guarantees—the survival of Cetus as a company." The firm netted $107.2 million on the sale, and Farley predicted that by 1985, when recombinant DNA and other biotechnologies would begin to pay handsomely, Cetus would still have $90 million left, all of it earning high interest. Just as the soaring of a company's stock in trading has no effect on the firm's treasury, so the plummeting of its shares has no impact, as long as the company never needs to return to the public for more. Further, his prediction about how much cash Cetus would have as its first products hit the market proved almost exactly right.

"The gentlemen on Wall Street keep saying how volatile the market is, how inflation does this and recession does that," Farley said. "We are out of that market, we can concentrate on achieving the goals that will yield the products that will justify everything we've claimed."

If there was at that time a difference in attitude between Genentech and Cetus toward its public financing, there is at least as great a difference in their relations with corporate partners.

Genentech prides itself on having sold few large blocks of its equity, all under very favorable conditions. Lubrizol, interested in diversifying away from lubricating products, paid $10 million for one million shares in 1979, then bought out Inco's original shares for another $10 million a year later to give it just over 20 percent of the firm. Later, Fluor Corporation bought about 4 percent for $9 million, but in return brought valuable knowledge to the partnership as a major builder of industrial fermentation plants.

Perkins feels the clear focus of Genentech will be especially important in the future, when the price of developing new products may increase sharply. Perkins notes that genetic engineering companies so far have made and marketed only products that the body naturally produces, albeit in purer form or increased dose. Hormones and antibodies are made by the body; thus interferon is a new drug but not a new creation, and monoclonal antibodies are very highly concentrated, but not novel antibodies. But somewhere in the future will come the unheard-of. For example, "There are something like thirteen types of interferon produced in the body, and they are all long, complex molecules," Perkins says. "There's evidence that when you recombine them—match the head of one, tail of another—you can get something hundreds of times more efficacious. That may be in our future."

But, "As soon as you enter this arena, the FDA aspect becomes much more complex. At that point you're dealing with the totally new. At present the only things you generally need to prove is efficacy and purity, because the chemical is already in the body. With all-new drugs you have to prove everything, from start to finish," and that represents a much longer development period and a much larger corporate commitment.

In the period after Cetus's public equity sale, the company was frequently criticized for spreading itself too thin. Farley, who was president then, countered that with its enormous financing and strategy of identifying strong partners to develop promising new ventures, Cetus represented the "largest critical mass of biotechnology in the world." Nevertheless, the strategy cost heavily in equity. Until 1977, Cetus had run on $5 million in venture capital. Then Standard Oil of Indiana (Amoco) came in with $10 million and got 22 percent of the company in return. Amoco, Chevron, and National Distillers, the three largest shareholders, owned 65 percent of the firm by 1983. Now the company's equity picture has changed dramatically. Amoco remains the largest shareholder, with 17.9 percent, but it is the only shareholder with over 5 percent of the company. According to a Cetus spokesman, shares are dispersed among institutional investors and individuals, and even when company officers' and directors' shares are lumped in with Amoco's, the total comes to just over 25 percent.

Cape points out that Cetus "had been in many areas, true, but that kind of diversification by other companies has been applauded." Today's Cetus is a "fully integrated pharmaceutical" company, as is Genentech, with a major line of concentration: cancer therapeutics. And Cetus made an impressive score in this home territory in 1985 when it was granted a patent for Interleukin-2, the anticancer drug that has shown such great promise in clinical trials on tumors that had been otherwise untreatable.

Cape also says that Cetus's somewhat high diversification became a cliché, used to explain unrelated events. For example, Farley had been very interested in Cetus's development of genetic engineering methods for producing high fructose corn syrup (HFCS). Cape says now that HFCS seemed an excellent candidate for genetic engineering in that period because soft drink companies were using it in place of high-priced sugar. Then, "Sugar went from seventy cents a pound to ten cents a pound," Cape says. "It's not surprising that there's not much market today for HFCS." Further, "Chevron independently changed its mind, but I remember some gossip sheets predicting that Chevron's canceling of the HFCS contract would have dire financial consequences. We had $90 million in the bank then."

Most importantly, Cape says, "In therapeutics, we are the only company to have no licensing agreements with anyone except Shell. We haven't tied up Japanese and European markets, for example, with licensing deals there. We have Eurocetus to do our own European marketing. We are as diversified as we were in 1980, but in all areas except human therapeutics we are diversified with someone else's money. Our perception is that if we are to do worldwide therapeutics by ourselves, our resources are barely adequate, and we must give our all to that goal."

If Cetus and Genentech share this concentration of resources as a major part of their game plan, they share another characteristic that is the envy of many companies: the ability to attract and keep some of the best young molecular biologists in fields related to their corporate interests. Here, Genentech certainly had the early lead, building a cadre of top scientists including David Goeddel, who continued publishing and carrying out front-rank science in a corporate environment—something which had never been done in the life sciences. Now Cetus can boast of such leading scientists as Shing Chang, a world expert on the bacterium *B. subtilis,* which many people see as the eventual successor to *E. coli* as the bacterium of choice in industrial genetic engineering.

Cape: "We keep track of those who leave here, to learn why people do go to other jobs. Those few who have left from the senior-scientist level have gone on to such positions as department chairmanships at large universities. Their forward momentum, in other words, has neither been lost nor diminished by their being here. They've maintained and even increased their peer recognition with publication and peer-group meetings."

And on the other end of the spectrum, Cetus was proving its ability to attract top young scientists, as Mike Kriegler moved from cloning tumor necrosis factor at Philadelphia's Fox Chase Cancer Institute to being part of Cetus's anticancer therapeutics development team. Clinical trials between Cetus and Fox Chase were underway by the end of 1986. Kriegler loves it. "I have complete freedom—if anything, more than I did before."

Freedom for the scientists, restraint on new product ventures, an unusual combination, perhaps, but one that appears to work. Bob Johnston also points to the latter as vital to corporate survival. The board chairman of both Genex and Cytogen, which he also founded, Johnston wants to see Genex's focus tighten, praises Cytogen for its pure concentration. The latter company, founded to exploit monoclonal antibody technology, is now concentrating on the linkages needed to hook diagnostic molecules such as radioactive imaging elements onto the antibodies.

"A lot of the failures in monoclonal antibody use occur because the linkages don't hold," he said. "We are developing some that we believe will hold." And that will mean that later the firm may have proprietary linkages to hold the antibodies to therapeutic molecules, such as anticancer drugs, a far more important future use of monoclonal antibodies.

 It was because of Genentech's refusal to divert from its strategic goals even for so promising a technology as monoclonal antibodies that Kleiner, Perkins, Caufield, and Byers founded Hybritech, to do what Genentech would not. Partner Brook Byers became Hybritech's president. In 1985, in a landmark some say is at least as important as Genentech's public sale, Eli Lilly bought Hybritech—for a stunning $300 million. As Collins notes, "A valuation like that by a group as savvy as Eli Lilly put a valuation on the entire industry. It said that Lilly was putting its energy, its muscle, here, and here is where you're going to see action." At the same time, Bristol Myers bought Genetics Systems, one of the oldest genetic engineering firms, for just about the same price. If Genentech's stock sale marked the birth of the industry, the twin purchases by two such profit-minded corporations marked its coming of age less than a decade later.

Weighing In

Analyzing the character and "weight" of the new industry through sales forecasts, actual sales, and real earnings, as analysts do, is an important way of reading the future, but Perkins, Cape, Johnston, and other industry executives feel that in many ways the industry is not being weighed properly as a high-technology enterprise. Such firms must be seen as people-intensive. The inability to judge what people will do in a new technology ought to lead to a certain forbearance: There is, they say, a need to withhold judgment and to keep working.

The first goal of the industry is to replace existing technology: for that, no new markets need be created—the firms need to link up with companies that have created markets, and they have done so.

Perkins: "The best way to look at this industry is to look at the potential market. Consider the possibilities for that market. Then ask: Can Genentech have an impact on that market? How much? If you look at the *whole field* like that, and you start adding up the potential profits, you get a very big number—a *very* big number."

For example, Perkins said, drugs have a very high profit margin if the developing company hits. "We believe we have a good chance of hitting." But this is a risk industry. "It isn't an area where you can know you're going to make this much money on this date. It's not a Southern Pacific bond issue, but there is business going on here."

The profit *potential* also seems much larger for genetic engineering than for other new industries, because industrial operations whose products are also made in nature have analogs carried out via enzymes. These are candidates for genetic engineering. If only a small percentage of the possibilities hit as well as early ones have, the industry will be enormous, as will its impact on the average citizen.

Most companies' strategies involve bringing in enough revenue to survive until "the big one." Most do contract work for clients—as with Genentech's developing insulin for Lilly, or Biogen's similar project for Novo. Contract work does not, as Collins noted, provide for explosive growth or justify large valuations, but it isn't usually intended to; it's intended to keep the firm afloat. One scientist said appreciatively, "Genentech has just about covered its research costs with contract work." That is important because it means the better genetic engineering firms may not need to return either to public or private financing sources until they have a specific project they want funded.

We liken the strategy needed for genefacture firms to that used by serious poker players: The startup money is the ante that gets you into the game. You shouldn't think of it as anything more than that. Even if it's in multiple millions of dollars, it just allows you to sit down and play. All the rest of the income represents the small hands you win while you're waiting, the hands that enable you to stay in the game. But you're waiting for the big play, and the trick is to be able to last; the odds say the longer you last, the better your chances of its coming. You have to *be* there when the big one comes, then you have to play it right. The major discovery—that billion-dollar bug that, say, gets 10 percent more oil out of the ground—that's what it's about. That's why you're in there.

And "there" means on the cutting edge. Dr. Morris Bell thinks most evaluations of genetic engineering have completely missed because they ignore the nature of the territory.* In a real sense, he is Collins' opposite number: "brought up on Wall Street," where

*Bell is the pseudonym of a scientist with unusual business expertise, who did not want to be quoted by name. He is not quoted elsewhere in this book.

his father has spent his life, he majored in Marxist economics, then "hit the Street" himself for several years. But he had always liked science, so he went back to school and became a molecular biologist. He now specializes in bacterial molecular biology at The Rockefeller University in Manhattan, and he also serves on the business board of a genetic engineering firm.

First, Bell says, let's assess the financial risk: "Are we comparing a speculative new business to a maker of widgets? You can't figure the two the same. How much does a corporate jet cost? Four to five million. That's how much big companies are sinking into any given genetic engineering firm, and they're getting a lot more for their money than a new jet. How much does a Broadway play cost to produce? Four to five million, and the odds on success are even worse, but you've got loads of people who just love to invest in 'em. Of course, the point is that if you want immediate cash profits, don't invest in biotechnology, invest in widgets."

Overvaluation? "Not at all, there's a real logic to most of the valuations. First, you need $5 million to $6 million to start up one of these companies; it's not wise to start with less. If all your equity goes to the investors [i.e., you give too much equity for a given price], there's no equity left for the scientists. If you didn't need to give the scientists equity to get them, the big drug companies would have gotten them long ago." Once the equity is parceled out and some is reserved for future science-hires, Bell says, the valuations frequently work themselves out by simple mathematics into the $40 million range.

BioTechnica's board chairman, John Hunt, points out one of the unique elements of the scientists' positions in the firm: "In a real sense, there are no genetic engineers. There are engineers in every other science, but here the work is carried out entirely by the top people in the field." That creates special management demands. "You're talking about a collection of very creative, brilliant people, and far more than even in a university, they must be able to work together."

Bell: "That's right, but an engineering class of scientists will develop here." And there will be another kind of shaking out: "I think there are already some of the top scientists in the business who'd rather be back in the lab, and at some point they'll move that way as others move out of the universities."

Walter Gilbert similarly believes that soon pure recombinant

DNA research will split from applied research, as has happened in every other science field, and the two branches will naturally develop in different ways.

Further, because the firms are so totally defined by the talents of their scientists, they represent a real resource *before* they've sold a product to other industries that support them. "The sources of information cost hundreds of millions of dollars to develop in universities," Bell said. "Making an investment in one of these firms is like plugging in to a terminal of the world's greatest computer. They're tapping into an incredible pool of knowledge. That's not true in the electronics industry—there is no such pool. If you want to know [computer] chips, you don't find a university scientist, you talk to the chip man at ABC computer. There's no precedent for this."

Bell, in fact, thinks gene-splicing companies should show they are taking some real risks *before* they become worthy investments. He pointed to one biotech support firm admired by King because of its traditional steady sale of quality products to other genetic engineering companies. "And that's exactly why it may always be a good company, but it will never make a huge hit," Bell said. "It will never go right out there to the edge and take a bunch of big risks on projects, some of which will never pay off."

But back to the widgets: "Biotechnology doesn't even come off badly by comparison. Let's say you're into widgets and you're optimistic, you see a 10 to 15 percent increase per year in demand, and you're well-positioned among other companies in widget manufacture. If that were the case, you'd expect to get fifteen to twenty times earnings for your stock. Now if you can see a cash flow of $30 million to $40 million you'd say that justified an investment of $600 million to $800 million." Looking at the whole genetic engineering industry, Bell says, that's right on target.

Bell points up a key distinction those practicing in the field say is not being made by most analysts. The near-term future of the *industry* is not bad, even though many individual companies may go down. "A lot of them will deserve to," Bell says. "When it comes to second-round financing, they'll either be very good, with a real 'score' they're hoping to finance, or very bad, with no products."

While many see growing threats in the field from foreign competition, Bell does not. The Europeans have tried to create their own industry with government sponsorship; but because that has meant no equity to scientists, they have had little luck. Biogen, for example, is a Netherlands registered company but operates out of Switzerland and Cambridge, Massachusetts, and does provide its top people with equity. It has no government ties, and as Gilbert noted, Biogen was created with a strong emphasis on control by its scientists.

Recent activities suggest that the United States needs to be concerned about foreign competition in genefacture. As mentioned earlier, the German government's investment of a billion marks ($300 million) into developing biotechnology, in a country with historic skills in fermentation technology and biology, suggests that even at this late date, the Germans could be tough competitors. But most American executives are primarily concerned with competition with the Japanese. Cetus's Cape believes that unless the United States recognizes this situation, the country soon will lose its number one rank in biotechnology to the Japanese. Already masters of fermentation technology, the Japanese have set dominance in biotechnology as a national goal. Soon, Cape fears, genefacture in America may follow steel, automobiles, and electronics into the second rank.

In a competitive world, that one country got an early lead in this powerful technology and now has challengers from the United Kingdom to Germany, from Japan to France, is a healthy sign, not a cause for lament. But we'd like to put these points about international competition in perspective, to understand the concerns of Cape and others. The right combination of science, business, and government played the major roles and luck very minor ones in America's speed off the blocks in recombinant DNA. So consider this sketch of the three players.

In science, basic research has been almost wholly carried out in universities, yet, especially since the close of World War II, carried out with federal dollars (Chapter 7 will cover the perennially uneasy relationship between universities and government). The federal government has pumped huge amounts of funding into science but, by and large, left the direction of research to the scientists themselves. Nowhere else has so large a government cash commitment been entrusted to peer review for disbursal—and the system has worked to a degree unimagined in other countries. U.S. biotechnology is where it is because basic researchers studied what interested and excited them. The government neither led them here with carrot on stick nor put up detours to head them elsewhere.

The business: American business innovation shows an interesting combination of freedom and restriction. On the one hand, through the diverse sources of venture capital, cash is available to embark on an enormous variety of projects in the United States; further, since this cash comes from so many different individuals and groups, no national consensus is needed to strike out in new directions. On the other hand, American business people describe their own game plan as short-term relative to those in other countries. In Japan, national goals may limit the kinds of undertakings at a given moment, but they also enable far longer range planning and deferral of gratification—that is, profit—than Americans are used to.

The government: In funding research, the federal government has played the role of midwife, assisting in the development of ideas and inventions to speed the transfer of technology from laboratory to marketplace. Only twice has the government set major national priorities for science: during World War II, in a variety of programs of which development of the nuclear bomb was paradigm, and during the 1960s in the race to put a man on the moon.

In sum: U.S. biotechnology in 1987 reflects the government's no-strings yet generous funding of basic research; it owes little if anything to the setting of national priorities. Now the government has sharply reduced that research commitment at the very moment when gaining the lead in biotechnology has become a national priority in countries that live by such goals. There lies the cause of disquiet for those watching the field.

Finding Your Way in the Gene Age

We have made a lot of points about science and the business of genetic engineering that may be useful in evaluating products and companies, for large and small investors or for research directors. We'll conclude by summarizing them.

WHO ARE THE COMPANY'S SCIENTISTS?

Genetic engineering is extremely difficult in almost all cases where it's useful, and this has remained true even as earlier "cutting edge" tricks have become widespread. And that fact leads to important conclusions. To put a simple gene into *E. coli* and just get expression was not difficult even in 1980. To maximize gene expression and overproduce protein was very tough. Now that's easy, but the field has rushed onward, and it is still true that few people have the bag of tricks needed to do the job to create tomorrow's products at a large scale.

How can you tell if a company has good scientists? The company's track record is the best single indicator. For example, Biogen, Cetus, and Genentech were among the first genefacture firms to complete genetic engineering projects, carrying them from the laboratory bench through, in some cases, industrial scale-up. Industry analysts based their early high marks on those projects. Now many other firms have completed similar work that stands up to critical examination. That is assurance that these companies have the technical capability of staying at the forefront.

Newer genetic engineering companies—which means many companies in this industry—haven't established records yet. But there's still a way to get a lead on their chances, by directly investigating their scientific staffs through information from a library or the company. What kind of publications have their leading scientists got to their credit? Are they or were they pioneers in university labs? Past pioneers tend to be future pioneers. Most important, if a firm's top scientists don't have a number of publications related to genetic engineering, that's an indication the firm lacks expertise.

Most companies boast advisory staffs of molecular biologists—well-known ones wherever possible—but in evaluating the staffs make sure they *also* have full-timers experienced in genetic engineering. No matter what abilities the advisory staffs provide, it is the full-timers who will directly affect the pace of research and development.

Pioneering is important in another respect. Developing new techniques puts a firm in a good patent position, and licensing of patents may turn out to be a large source of future income—much larger than for universities, for example, which do not seek high returns on licenses. Patent position may be especially important related to organisms other than *E. Coli,* and to fermentation scale-up. The newness of biotechnology patents certainly clouds their future, and there are quite a few heated patent fights going on right now (see Chapter 6). Nevertheless, the companies with the best scientists and fermentation engineers stand the best chance to get projects into commercial production and get defensible patents.

WHAT'S THEIR LINE?

Most of the new Gene Age companies came into existence between 1980 and 1982, when venture capital and other sources of financing were plentiful. By and large, they started as research and development firms, with only a vague business focus. As the first phase of research and development is now nearing completion, their business goals and early products should be fairly well defined.

How do you determine which companies will have commercially successful products, when most products have not reached the market? Sales and earnings, after all, can't be factored in here, so we must guess which products and businesses have a high probability of making money. We won't analyze specific products or firms, but we'll comment on pluses and minuses of some concepts.

Of the pharmaceuticals that dominated initial company development plans, interferon was the most publicized. Since research and development on many of these pharmaceuticals was done in the early 1980s, many products are in clinical trials and a few are on the market. They may change medicine in profound ways, but will they make money for the companies producing them? Let's consider some potential pitfalls between here and the marketplace.

The computer industry has no FDA. The high cost of clinical trials has already been noted. But on the positive side, the FDA has taken a reasonable view toward many products, requiring only that the manufacturer show that the product is pure, if it functions as a simple replacement for a natural substance. Whether a product is made by recombinant DNA should be of little concern to the investor. But the regulatory situation of the product and the estimated cost and time to regulatory approval should be of major concern.

Next, some of the more exciting and potentially profitable items are being developed by several companies. Who will get to market first? More importantly, will one company get a defensible patent position, completely blocking competitors from marketing? If that is not known, risk increases considerably. And that risk is further worsened by a backlog in processing biotechnology applications in the U.S. Patent Office; delays are running two or more years. That means that a company could find out two years down the R&D road that another firm has won a patent on the product or process of making it, or on an invention so similar that further marketing is effectively blocked.

A key component of "what's their line?" is "what's their mix?" That is, while focus is good, a company working on only one product could be asking for serious trouble. A lot can happen to block introduction of any given product: R&D fails, regulatory approval is denied, a competitor develops a better version, beats the company to market, or presents a superior marketing force. Finally, for new products without an existing market, acceptance may be very slow.

Having more than one product in the pipeline is a vital sign of corporate health. But why is everyone so uneasy about having *too many* under development? Because research and development efforts can be diffused until progress on all fronts slows dangerously. Most small companies do not have the financial muscle to carry out many lines of research, product development, and sales and marketing simultaneously—unless their technology and business expertise allow several products to fit into the same pipeline. Lack of focus can be fatal.

How has the company timed introduction of its various products? For example, agricultural products may be aimed at very large markets, but since plant genetic engineering is just beginning to show success in the *laboratory,* the road to market is a long and winding one. To actually make money selling seed to farmers, after the genetic engineering is successful a company still must grow whole plants producing seed from the engineered cells. The seed must be tested for one to three years in greenhouses and small-scale outdoor plots to see if the new plants behave as predicted. Then large-scale field trials are in order. Finally, enough seed must be produced through large-scale plantings to offer for sale to farmers. At every stage, regulatory hurdles may have to be cleared. Timetable: After the basic genetic engineering is completed and "victory" is sensed, four to seven years may be needed to make money on the product.

While many of the entrepreneurial companies have portfolios heavy with new-wave pharmaceutical or agricultural products, however, they also have other means of producing income over the short term. DNA probe diagnostics and genetic-defect testing represent two means of earning near-term revenues. Developing a DNA probe for a specific diagnostic test often costs little money and can be done in a year or two. Tests using these probes unfortunately still are not "user friendly" enough to be administered in a doctor's office, but they can be done in a relatively sophisticated clinical laboratory. Many firms are setting up their own clinical labs to carry out infectious-disease testing or genetic counseling.

Of major importance, diagnostic tests are not subject to as laborious a regulatory process as therapeutics. That's especially so if the test is relatively noninvasive, such as those requiring vaginal smears or small blood samples, *and* if the test is used to detect a disease that is usually not life-threatening, such as those for most venereal diseases, periodontal disease, and potential genetic diseases *before* conception. Diagnostic tests can be commercialized in two to three years, compared to four to ten years for many pharmaceutical and agricultural products.

Industrial chemicals and enzymes, while not offering the high added value of pharmaceuticals, do offer a way to near-term revenues, because they generally are subject to little if any federal regulation. An appealing strategy: A company might enter a market with an industrial product it has made by "classical" means. This allows earlier entry into the market and produces needed cash flow, offers experience in the marketplace, and paves the way for eventual replacement of the company's own product with a "Gene Age" version. Cetus is following a similar strategy in producing methotrexate, a conventional cancer chemotherapy drug. Chairman Cape noted, "We want the name of Cetus known to oncologists before Interleukin-2 is marketed."

Finally, of course, contract research for large companies offers the most common method of short-term revenue generation. The larger, wealthier company is in a better position to assume the risk of R&D, but not surprisingly, he who assumes the risk takes a large share of the reward. As a result, while the small company might get a modest future royalty on the product under development, it is really often trading a piece of its future for current income and reduced risk.

Terms of such contracts are vital. An arrangement that gives the big company all future rights to the developed technology, for example, is not a good one for the small company. On the other hand, one that gives away partial rights to just one class of products, or for one geographical area, can be beneficial to both companies. That can allow the "start-up" to develop its technology base cash- and risk-free and secure its own market niche or territory.

The quality of contract clients is one measure of a small company's technical abilities. Contracts with companies well known and respected for research in their own right provide strong evidence of the small company's technical skills, because the large firms can be considered good judges of those skills. But such confidence should not be taken as judgment of the small firm's business acumen—the large firm may be interested only in technical results and not the small company's odds of survival as a business. For that type of evidence, look to equity investments by larger firms, generally a better indicator of faith in business skills.

Finally, what are the projected dollar sales of a company's product line? Naturally, there will be a lot of guesswork here until many more products reach the market. Somewhat arbitrarily, "big products" are generally considered those for which potential total annual sales by all makers is over $100 million, medium-sized products are those with potential sales between $25 million and $100 million, and small products are those with total sales below $25 million.

For example, some new-wave pharmaceuticals, especially anti-cancer or heart drugs, may be big products. Most diagnostics fall in the small to medium range, but a few might hit the big-product category if they are used widely. Agricultural products that affect large-volume crops, such as corn, soybeans, and wheat, can be big products. And commodity chemicals would represent big products if they could be made competitively by genefacture. It's doubtful that genetic engineering can compete with petrochemical processes right now. Specialty chemicals are mostly in the medium to small size range.

How can the potential investor estimate the value of a product to a company's stock worth? Here's one way value is often estimated—although in applying any method to new products, "guesstimated" is probably the better term. First, estimate the total market for the product. Often, the company's literature will give you that figure. If you're skeptical of the company's projection, brokerage houses often have information on total product markets. And the U.S. Commerce Department publishes an annual survey of manufacture that puts a value on major world products. Sometimes your own common sense can lead to a reasonable rough estimate. For example, it's easy to come up with a figure for the toothpaste market in the United States of between $500 million and $1 billion a year. There are 230 million people in the country. If we figure that two thirds of these people use three tubes of toothpaste a year, that means a total sale of about 460 million tubes annually. A check at the supermarket shows the cost at about $2 per tube, at which sales would be $920 million.

Next, estimate what you believe the company's share of the market will be. For a new product with no existing market, figuring 20 percent market share may take into account competition and incomplete market penetration. So for a product with a $100 million total market, the company would be expected to realize $20 million in sales. Making such calculations on a new product is plainly full of hazards; and even making such estimates for replacement products has its own pitfalls, not the least of which is the probable need for it to displace whatever is already in use. Again, brokerage analysts often have a good handle on market-share calculations. Call any broker and ask.

The next key number to calculate is after-tax profit. For high-tech products, we believe a reasonable number to be 15 percent of sales. So for the previous example, $3 million profit per year could be expected from the $20 million in sales. Now, how might that affect the company's stock price? We'll guesstimate again. For a growing company, a stock price-to-earnings ratio (P/E) of 15 is reasonably conservative. That means that $3 million in annual profit (the earnings contribution this product is making) will contribute $45 million to the company's value. If the firm has issued 10 million shares of common stock, the contribution per share is $45 million/10 million = $4.50 per share.

Now, let's say four products are coming down the company pipeline with similar value. You might peg the company's future stock value at 4 × $4.50 = $18 per share. If you're buying at $2 a share and expect the four products to reach full market penetration in five years, you're probably getting a good deal. Most companies, however, are already trading somewhere near their future value. The stock market often values on future worth, and certainly has done so on most biotechnology companies, which have little present earnings.

Feel uncomfortable with this analysis? Concerned with the risk that a particular company faces? We are, too, but it is all anyone has to work with, including stock market analysts. While professional analysts have somewhat better numbers to work with, they are still, to a large extent, guessing. Evaluation of a company is full of uncertainties, uncertainties that reflect the risks entrepreneurs live with every day. They keep plunging ahead, lured by the pot of gold at the end of the rainbow—speaking of which . . .

ow Do We Get There from Here?

Most observers still believe that the U.S. Office of Technology Assessment's $27 billion business forecast stands up, that that is the value of products that could be displaced by genefacture by the turn of the century. Some believe that it will take years longer. Given the surprisingly fast rate of technology development in such slow areas as plant genetic engineering, for example, we side with the majority and believe that timeline to be realistic.

Today sales of genetically engineered products are quite modest, with 1985 sales totaling about $47 million, according to Shearson Lehman Brothers. How, then, will the firms in the business get to that golden future? The key, naturally enough, is to retain enough money to stay in the game—to avoid going broke. So in analyzing a company's worth, we need to calculate its staying power, and that means its financial condition: Does it have the means to carry on planned R&D, and marketing and sales for the time necessary to sell enough to support itself? If so, it likely will get there.

Let's compile a financial profile for an "average" genefacture company. We've selected five actual companies, all among the new entrepreneurial firms, from which to develop the average. None of the firms is older than eight years, and all were formed to exploit the new genetic engineering technology.

The selection was somewhat arbitrary, except that we wanted both larger and smaller companies (with the understanding that, compared to industry giants, these are all small companies). We took these: Genentech, Biogen, California Biotechnology, BioTechnica, and Cambridge Bioscience. Size range: from about 70 employees to 1,000. Since we're trying to come up with a profile for an average company, we scaled the dollars for each company to one with 150 employees, and that's probably a typical size; Genentech is far larger than that.

Here are the dollars, scaled and averaged, for the company in 1985:

Financial Category	Millions of Dollars
Cash and other funds readily convertible to cash	$ 17.6
Revenues	8.6
Expenses	11.9
Net income	(3.3)

Figures for this table were taken from the companies' annual reports and from data summarized in Shearson Lehman Brothers' excellent review, "Biotechnology: From Lab to Marketplace, an Emerging Industry."

Let's look at net income—revenue minus expenses. The average company in 1985 was losing $3.3 million a year. Revenues, however, were a noteworthy $8.6 million a year, and were generated mostly from contract research with larger companies, from the R&D partnerships we referred to, and from interest on cash or equivalents in the bank or any other interest-bearing instruments.

Interest is a significant source of revenue now, since the average company has $17.6 million in the bank. At 8 percent, that provides $1.4 million per year revenue. Thus, the average company is on its way to break-even, but it is still burning cash. One to three new contracts or product sales in the $10 million to $20 million range could make up the $3.3 million a year deficit to bring the average firm to the break-even point.

Can Average Company survive to break-even—that is, to the point of being self-supporting? With $17.6 million in the bank and a "burn rate" of $3.3 million a year, it obviously now has five years to become self-supporting (17.6/3.3); further, that cash must be used entirely to finance operations, not for expansion of operations or large capital expenditures.

Five years should be more than sufficient for most companies not only to break even but to reach profitability from earnings on sales. Those that are tight on cash and do not have a five-year or longer margin of safety were seeking more capital in 1986. Many were successful, as biotech investments returned to popularity before the stock market setback in September.

Outlook: favorable.

hat's Down the Road?

Survival is nice, but overwhelming success is nicer. Several long-term projects in genetic engineering have the potential of being billion-dollar winners—if they can be shown to be economically feasible. Such projects are found in the areas of nitrogen fixation in plants, energy production from waste biomass, and basic feedstock manufacture for the chemical industry. Forecasting the economics for these projects is torture. For all these products, the economics depend on the future price and availability of petroleum, and for some there is a question of whether they can be carried out at all.

One must be wary of a company that speaks of its future only in terms of these kinds of high-risk projects. But any forward-looking firm ought to have a few, far-out "great notions" in its project bag, to keep alive the chance to become one of those that will be remembered for opening up the Gene Age.

NOTES

[1] *Wall Street Journal,* "Ex-Convict and Microbiologists Join Forces to Ride the Genetic-Engineering Wave," March 4, 1981.

[2] *Time* magazine cover story, March 9, 1981.

6 CHIMERAS

CHIMERA *(ki-me-ra): Head of lion, body of goat, tail of dragon. Fire breather.* The original chimera was a whimsical piece of work, whether of nature or of a fertile Greek imagination. But there was no whimsy in Patent No. 4,237,224 as it was issued December 2, 1980, announcing itself simply as:

PROCESS FOR PRODUCING
BIOLOGICALLY FUNCTIONAL
MOLECULAR CHIMERAS

The Stanley Cohen–Herbert Boyer process patent details how to make a "plasmid chimera," so named because of the unnatural relationship of its parts: plasmid DNA conjoined with the gene coding for human insulin, stitched into a ring, and neatly inserted into an *E. Coli* bacterium, thus creating a novel life-form.

In the few years since 1980, hundreds of patents have been applied for covering a wide range of microbes, some produced by forced mutation, a traditional method of deriving "novel life-forms," others via genetic engineering. So far, that has been a positive sign of the burgeoning industry. Several patent fights have begun as well, however, and there is some concern that if patent suits get out of control, companies could drain their coffers for court battles over their products.

Patents are grants for invention. They grant temporary ownership rights to those who use scientific facts to twist natural elements into new shapes, press them to new function; they presume there are many new things under the sun. Discoveries of principles and natural laws are just the opposite: the elucidation of something that has always been there, regardless of how previously unknown, some part of all that is presumably eternal under the sun.

That boundary between the novel and the natural, often difficult to discern, was the site of the most major legal battle so far in genetics, one that ended in mid-1980, when the U.S. Supreme Court decided that whether an invention was alive or not played no role in its patentability. The decision involved not the critical Cohen-Boyer discoveries that put recombinant DNA "on the map," but a relatively obscure microbe created by General Electric scientist Ananda M. Chakrabarty.

Different natural varieties of microbes are known for their ability to digest various simple hydrocarbons, but none can eat and transform the complex variety that makes up crude oil. Chakrabarty used a combination of naturally occurring plasmids from different strains of microbes, an older process that is only genetic engineering in the broad sense, to develop a strain that could eat the whole thing. Even though neither the scientist nor GE believes that microbe is ready for commercial development, the Court's grant of "ownership" of this microscopic living thing, by a 5–4 ruling, marked the watershed. That was June 16, 1980, but long before the Court clarified the issue, much controversy and no little confusion had been injected into the dispute.

Some writers had been wondering if the patentability of life suggested terrible new forms of slavery in which cloned humans would march about with patent numbers holding them in bondage down through their generations. Others suggested that science had been given new license to create monsters and conjured visions that Chief Justice Warren Burger dismissed as "a gruesome parade of horribles." And there were still others who worried that the profitability of patented microbes would lead to secrecy in the world's leading biological laboratories, a trading of scientific progress for royalties.

Unraveling the implications of the Chakrabarty decision requires first coming to grips with what the ruling did *not* do. Patent lawyers, who fought for the ruling as a friend of the court, generally are agreed that the importance has been greatly overblown. To begin with, a patent was granted on a living organism to Louis Pasteur in 1872 for a strain of *naturally occurring* yeast, and that may not even have been the first such grant but one that stands out because of Pasteur's fame.

Bruce Collins, an attorney who wrote for the American Patent Law Association in its filing as friend of the court,[1] explains: "The Patent Office then probably treated the yeast as simply an object of commerce and did not specifically think of it as a living thing." By the middle of the twentieth century, microorganisms were described as part of a large number of patent applications involving industrial processes, Collins pointed out—most notably fermentation processes for the production of antibiotics and vitamins. But in those cases the invention involved a new *method of using* the microorganism, not the microorganism itself.

Patents on different kinds of inventions are available: process patents may be obtained on new methods used to make something old or new, and product patents are given on the "new things" themselves. Process patents involving microbes as integral parts had always been available. Product patents had not been—but neither were they needed. No new living things had been created. Or had they? What about hybrid plants? Addressing just that narrower issue in 1930, Congress added plants—very much living things—to the list of novel material that may be patented. The Plant Patent Act made no reference to animals or to microorganisms. One question Chakrabarty brought before the Supreme Court was whether such protection ought to be given to *any* novel living strain. Objectors said that had Congress intended non-plant life to be covered, it would have so specified.

Chakrabarty's attorneys argued otherwise, and the Court majority agreed with their interpretation. When any application is made to patent an invention, a full, detailed description must be filed explaining the makeup and function of the "device." The only reason plants required specific congressional action was because providing a complete description of their structure would prove impossible, and proponents argued that the fact that the plants were alive simply had no bearing on the act.

Dissenting Justice William Brennan argued that Congress *had* foreseen the question of non-plant patent rights in 1930 but had not resolved it, and therefore the Court should not. The question therefore was a narrow one. (A little later we'll look at an unusual twist in plant patenting that resulted from the Chakrabarty decision.)

But to Collins and others in the field, a ruling against Chakrabarty would have gone against the intent of the patent law; a ruling in his favor more or less maintained the status quo. William O'Neill, a consultant to Stanford on the Cohen-Boyer patent license, said Chakrabarty "would have been news had it gone the other way. Then it would have been a complete subversion of the intent of the patent law, whereas this is just a logical extension of it."

Cetus's Peter Farley, whose firm also acted as friend of the court, said, "Our work in no way slowed in anticipation of the decision, nor did we anticipate changes in our work depending on which way it went."

Despite "exaggerated media interpretation of the results," Farley said, "In a sense the positive impact was in the Court's bringing genetic engineering as a commercial enterprise to the attention of the entire country."

The patent issued to Stanley Cohen and Herbert Boyer in December 1980 remains the most interesting one in the field, and the progress of its licensing agreement, developed by Stanford University, parallels the growth of the genetic engineering industry. That process patent involves the prototype tools of genetic engineering: the plasmid rings of DNA and the variety of cutting and sticking enzymes Cohen and Boyer used in their 1973 experiments. It could have been issued had the Supreme Court ruled against Chakrabarty's product patent application, because process patents have been issued for years.

But there is something very special about that Cohen-Boyer process patent. In some ways it may be the "Spirit of St. Louis" of genetic engineering—not born of the first recombinant DNA experiment, not even a patent on a "novel life-form," but a patent on the elegant bug-making process that electrified the scientific and business world to the possibilities of rDNA.

Despite the importance of the claim on the basic engineering process, patent lawyer Collins believes additional Cohen-Boyer claims concerning *production* of protein from foreign organisms will have an even greater impact. The genetic engineering process generally would be practiced just once in the course of developing a new microorganism; in addition, it might be used on a university level, where infringement charges would not be likely. But once the microorganism is made, then a firm would have to use these additional Cohen-Boyer techniques to actually produce protein, which in theory, Collins says, "covers virtually every application of rDNA at the moneymaking level."

It was not until August 1984, ten years after application, that the U.S. Patent Office granted Cohen and Boyer their claim on the actual novel microbe they had genetically engineered. The development of patent rights in general and court fights to defend them have become key factors in the progress of genefacture.

Developing and promoting the license under which Stanford and the University of California earn royalties on the twin Cohen-Boyer patents was largely the work of Niels Reimers, head of the Office of Technology Licensing there, and Andrew Barnes, an associate in the office. Barnes has since left Stanford. If the patent is the Spirit of St. Louis, Barnes was the "barnstormer" who showed it off to the world. Stanford developed its licensing plan to be a model of university–industry relations. As with most university licenses, it was not designed to make a fortune, but to get maximum exposure for the patent, reduce the chance of infringement fights, and bring a modest income as efficiently as possible.

Reimers and consultant William P. O'Neill wanted the Cohen-Boyer license to bring in supplemental research money for Stanford and the University of California, but at rates low enough that all interested firms could afford to sign up rather than fight in court or evade payments. Once the terms were worked out, it was Barnes's job to sell the idea, and he traveled throughout the United States, Europe, and Japan.

Reimers explained, "This wasn't just another patent. It was the central patent to genetic engineering." Because for now virtually all rDNA work requires use of that patent, "We wanted this to be the prototype for future patent licenses." Cohen and Boyer refused all income from their discoveries, normally 50 percent of the royalties, and that enhanced the potential for the universities' income.

The minimum license fee is $10,000 a year for all companies signed up, but with the aid of major firms in the field, O'Neill and Reimers devised an elaborate credit system to encourage early sign-ups—and that part of the plan paid off well. By December 15, 1981, the deadline for getting credits, seventy-three companies had applied for licenses, meaning a guaranteed $720,000 a year for the universities. Reimers had predicted Stanford would sign forty to fifty, while a more optimistic Barnes had expected fifty to sixty.

In five years, the Cohen-Boyer patents have brought $5 million to the universities, and eighty companies now have license agreements. Initially, Stanford had charged royalties on a sliding scale based on millions of dollars in products sales. Then in late 1986, the university upped the ante, now charging a flat 1 percent on sales of end products developed under the license, the most common way Stanford gets royalties, although the entire system is complex and subtle. The new rates apply only to new signers of the agreement, according to Reimers; early signers still pay based on the old sliding scale, meaning they generally pay less.

The importance of the large infusion of cash to Stanford was that it provided a sizable "war chest" for court challenges on the patent or action against infringers, and that may have deterred court fights. The only challenge to Stanford's agreement came when Cetus decided to drop its license, but the company later rejoined the pact. Many in the field believe the Cohen-Boyer patents might be successfully challenged in court, but such actions are so costly that a company would have to know it would save a fortune in licensing fees by winning—and to feel very sure of victory.

That could be important now, because the Cohen-Boyer patent may begin paying off next year with the first industrial "boomer" of genefacture—Tissue Plasminogen Activator. And how Stanford and Genentech stand to fare in selling the product offers a fascinating insight into the developing industry.

Remarkable first-year sales of $450 million have been forecast by Floyd Grolle, market research manager in Stanford's Technology Licensing. FDA approval for general use of the anticoagulant TPA was expected by early 1988. Grolle estimated that first-year revenues would go almost entirely to Genentech. If that forecast holds, the California universities would get royalties of about one-half of 1 percent, or $2.25 million, because Genentech was an early licensee; further complicating the revenue picture, Stanford gives royalty "credits" of $50,000 for each $10,000 a licensee has paid into the fund. But clearly the universities will get between $1 million and $2 million in that first year alone—if Grolle's forecast is correct.

The forecast was based on TPA's use for 20 percent of the

applicable heart attack patients in the first year at an expected charge to the hospital of $1,500 per patient treated. Interestingly, that is also considerably cheaper than the $2,000 cost of urokinase, the product currently used to treat patients. Reimers noted that urokinase must be purified from urine, and it takes the equivalent of one person's urine output for two years to gather enough urokinase for one patient's treatment.

By the second year, Grolle forecasts that 40 percent of all applicable patients will get TPA, but the price will drop to around $1,200 per treatment, which would mean $720 million in annual sales. By the third year with 60 percent use and a price drop to $1,000 a treatment, $600 million would be realized in sales.

Now comes the tricky part: No less than thirty-one corporate alliances, generally of pharmaceutical houses and biotech firms, are poised to compete with Genentech to produce TPA. Roughly half of those Reimers read from his list are *also* Cohen-Boyer licensees. Many of the others are foreign and don't need to sign up with Stanford. That intense competition could cut the cost of TPA drastically, reduce total sales revenues proportionally and, splitting the total among many players, sharply reduce each firm's share. In that case, the universities probably would fare well—they will collect royalties from all sales by licensees.

Or TPA could lead to the "Lebanization" of genefacture, a mutually destructive situation in which dozens of companies living in an uneasy truce pour large amounts of their resources into defending patents involved in the production of the drug, hoping to reap a larger share of the market. That would be an extreme outcome. Nevertheless, the corporate rush to get a personal stamp on TPA production certainly indicates the growing importance of patents to companies in genefacture.

Patent lawyer Bert Rowland, who filed the Cohen-Boyer application in 1974, sees a lot more patents in the future of genetic engineering, unless the field itself inexplicably dries up. "If the whole thing goes away, there will be no patent activity, but that's the only thing that will stop it." Against those who believe infringements will make trade secrets the "protection of choice," he noted that antibiotics offer the same chances for patent theft, yet most of them have been patented and have brought large profits to the companies that developed them. Further, in most cases, fights against infringements have been successful.

There is a major difference in production of antibiotics and some genefactured products. Generally the *end product* is patented on a new antibiotic, so infringement can be detected on the open shelves of the drugstore. Genetic engineering often deals with known end products, to be manufactured using newly devised methods or microbes, and they operate at the far more secret production level, where infringement might be hard to detect. But there is a major similarity as well. So far, the major products of genefacture have been new pharmaceuticals, and the pharmaceutical industry is comfortable with patent protection. That may help to explain why the early years of the industry have been dominated by patent application rather than trade secrecy, and the difference is an important one.

Patents are not entitlements to secrecy—quite the opposite. They offer inventors a limited monopoly on their work, usually seventeen years from the date of issuance. In return, the inventor must file a description of the invention so complete that anyone reasonably knowledgeable in the field could duplicate the procedure. Because plants and microorganisms can never be exhaustively described, a companion procedure many years old allowed for deposit of actual specimens in culture banks, in addition to the descriptions.

A trade secret is the perfect contrast to a patent, and the formula for Coca-Cola is a perfect example of such a secret. Reportedly that formula is known to only five people in the world; however, if others were to figure out the formula, they could make as much of "the real thing" as they liked, without paying the firm anything. If Coca-Cola were patented, anyone could learn the formula, but for the duration of the patent would have to pay royalties to use it commercially. Why use trade secrets? Because many companies believe it is easier to defend their secrecy than to defend patents against infringement.

For example, suppose a company patents a strain of *E. coli* that produces insulin better than any other. Another lab follows the procedure, obtains the strain, but classifies this process and microbe—actually stolen—as its own "trade secret." Noted one patent lawyer, "You may have a terrible time trying to prove that the strain you believe they are using is really the one you've patented." The patented strain might have to leave a telltale sign in the product for infringement to be proved.

Massive systematic infringement is especially of concern to European companies, which are afraid that patented microbes will simply be "stolen" and used in other countries in which there are no patent laws or in which discovering and proving infringement could be virtually impossible.

To Reimers, the *real* importance of patents is to prevent secrecy from coming to dominate commercial rDNA and strangling exchange of information, just as many scientists fear. And interestingly, there probably would have been no Cohen-Boyer patent had it not been for Reimers' persuasion. In 1974, with the deadline for filing application just weeks away, Cohen was set against the idea because he believed colleagues would see it as his attempt to monopolize his findings, Reimers recalled.

"The great debate over imposing a moratorium on rDNA research was reaching its height," he said. Cohen and Stanford's Paul Berg were among the leaders in the fight for the moratorium. They believed an application at that moment would seem part of a strategy to keep control of rDNA development. "But the main thing to Stan was that he wanted the process to be used widely," Reimers recalls, "and he was afraid patenting would limit its application."

Reimers changed Cohen's mind by recounting an historical incident. Oddly enough, the story appears to be apocryphal in part, but it was true as far as the major points Reimers wanted to make:

In 1928, the story goes, Alexander Fleming published his discovery of penicillin, indicating it might have curative benefits against bacteria, but he refused to file for a patent in that spirit of sharing knowledge that Cohen now wished to embrace. Drug companies, however, refused to work on developing penicillin for fear that their research and development money would be wasted, because competitors would simply duplicate their results for nothing.

The result was medically disastrous. Penicillin was not used until World War II—a dozen years after it might have been available—and it was then developed commercially by an American, and made its appearance on the battlefield as a *secret* of the Allies.

The source of Reimers' story was Betsy Ancker-Johnson, who he recalled had told it to a congressional committee some years before when she was Assistant Secretary of Commerce. Mrs. Ancker-Johnson and a former aide told us in turn that they had gotten the story from a book entitled *Miracle Drug: The Inner History of Penicillin.*[2] As with any story passed through many hands, this one had lost key elements and suffered some revisions, but the major relevant points were right:

Author David Masters recounted Fleming's discovery, and noted that penicillin was a laboratory curiosity known to have some antibacterial properties but—interestingly, much like interferon—was for years impossible to produce in quantities large enough to study. A few dogged researchers picked up the puzzle one after another for twelve years, trying to learn its properties, delayed by the difficulties inherent in the investigation (not by lack of *protection,* at this stage). Then, just as Britain appeared to be losing the war with Germany, the most dogged of all the researchers, Oxford Professor Howard B. Florey, made enough of the drug for him and his colleagues to realize it was not just a scientific curiosity but a miracle cure for many deadly bacterial infections. The British scientists were determined to see it brought to commercial production, at no benefit to themselves and for the good of all mankind—and they did so, eventually.

Professor Florey desperately pressed British drug firms to make large quantities for clinical trials, but he found the fermenters bogged down in the war effort and constantly threatened by bombs.

But, author Masters pointed out, another concern may also have been operating: many firms may have feared that as they attempted to scale-up natural penicillin for production, someone would discover a way to synthesize it and put them out of business. That had happened more than once before. As a matter of fact, a patent by Florey *would have* quieted this fear, but that does not appear to have been Florey's concern. He and a colleague flew to America and persuaded the U.S. Department of Agriculture to make the drug in large enough quantities for clinical tests. But at this point the story recounted by Masters contains a great irony apparently not known to Reimers: an American USDA scientist filed for a patent on the corn-steep liquor method he had developed for batch-fermenting penicillin. That patent was held by the USDA in this country, so Americans enjoyed the fruits of their tax-sponsored research at no cost. But that scientist, as was customary, got foreign patent rights *himself;* all the British, including Florey and Fleming, had to pay royalties to the American who had commercialized their discovery, while they got no financial reward. Fleming, Florey, and Ernest B. Chain, who also did crucial work on penicillin, won the Nobel prize in 1945 for the difficult discovery and development, but Fleming noted the patent-rights issue with no small amount of disappointment in a speech in New York that year.

As with many such old stories, this one reverberates hauntingly through the current discussions of university-government cooperation, as we'll see in the next chapter.

Confronted by Reimers' warning on the danger to the public in his not seeking patent protection, Cohen relented, and Stanford filed one week before the expiration date in November 1974. But the road ahead was rough. Reimers recalled that Cohen was standing in a crush of scientists at the rear of a conference as the moratorium was being debated at the Asilomar Center in California, when rumor of the Stanford application spread. The conference speaker spotted Cohen and asked him to step forward and explain. Cohen told Reimers he then took "one of the longest walks in my life" to the podium.

Granted that the universities will share the income from the patents, what will even several million dollars mean to major research institutions? Little and much.

Stanford's annual research budget is $150 million, which places that private university far below the vast, public University of California system, which conducts $500 million in research every year. Roger Ditzel, UC patent administrator, noted that roughly 10 percent of all university research in the United States is conducted in a branch of the University of California. Patent income has doubled in the past few years, but none of that income is from genetic engineering. "We hope the Cohen-Boyer patent pays off, but let's say we're not spending the money before we get it."

But Ditzel also pointed out that because the function of licensing is getting products to market, UC officials determined there should be a greater commitment to that goal. "I hope our patent income develops *because* we're doing a better job of transferring technology," Ditzel said. "University investigations are not designed for the marketplace. Good old industrial sweat and toil are needed for that. We're talking about getting companies to invest a heck of a lot of risk capital in embryonic products. With a patent license, we assume they can reduce that risk a little."

Further, even if patent income is small, it comes without strings attached, a boon to research administrators. Both Ditzel and Reimers point out that these unrestricted funds are invaluable in keeping departments running smoothly.

Stephen Atkinson, executive secretary of the Harvard Committee on Patents and Copyrights, put it this way: "Unrestricted dollars are simply worth more than other kinds of money." Such income can be spread around at the discretion of administrators, unlike funds for sponsored research. The latter, whether given by the government or industry, must be painstakingly accounted for.

Translated into human terms, that impact is considerable. Reportedly, one American university realizes enough income from a single genetic engineering patent to support two postdoctoral fellows and one graduate student, and that before the licensee has brought its product to market.

Further, as genetic engineering develops from a few basic principles to more specific, novel sets of procedures and more unique microbes, patents should become ever easier to defend. In the case of a patent claim, the more universal the principle underlying it is, the less the discoverers can call themselves inventors. Atkinson predicted that, "As differentiation occurs in the field, applications of genetic engineering in medicine will become very distinct from those in industrial enzymology, and those distinct from energy development."

So far that has proved to be the case with genetic engineering patents, and a recent court decision suggests that the government's attitude toward them is becoming increasingly permissive. The case involved a challenge by Monoclonal Antibodies, Inc., to a patent held by Hybritech. This was the first major case challenging the *validity* of a patent rather than which company ought to hold it, so the outcome was seen as a reflection of the federal attitude toward genefacture patents. One patent attorney said he found Hybritech's patent "of marginal validity. It was certainly *arguable* that the process described was obvious." What is obvious, like what is natural, is usable by anyone and cannot be patented. Nevertheless, the U.S. Court of Appeals for the Federal Circuit (CAFC) reversed a lower court ruling and upheld Hybritech's patent. As the major court for determining patent policy, CAFC seems to be leaning toward patentability of genefacture's processes and products, as it does in other fields.

On the other hand, because patents are so expensive to defend—running routinely in the hundreds of thousands of dollars—

industry watchers see such court tests as a mark of a company's belief in a product's value. Notice how these legal disputes coincide with other evaluations of genetically engineered substances: Biogen and Hoffman La Roche went to court in 1984 over the former's European patent on alpha interferon; Genentech announced it had won a British patent on Tissue Plasminogen Activator and lawsuits from other U.S. makers challenging the patent are expected, and Cetus and Roche went to court over Interleukin-2. In mid-1986 an international dispute broke out between two of the most prestigious research organizations in the world. France's Institut Pasteur sued the National Cancer Institute over the latter's exclusive royalties to several U.S. marketed AIDS test kits, based on the virus Americans call HTLV-III. The French say they found the virus they call LAV, provided the Americans with the sample they grew up and renamed, and are entitled to royalties. Sales of the test kits: $40 million a year and climbing.

The potential value of some genefactured products together with the pharmaceutical industries' penchant for patent protection seems to be the dual force behind the frenzy of patent activity in the field. But the result has been a tangle in the Patent Office: a backlog of more than two years with a recent estimate of 2,600 patents pending in the biotechnology field. According to a 1984 article in *Chemical & Engineering News,*[3] the array of applicants is as interesting as the numbers: of 846 biotech patents assigned in the first nine months of the year, 40 percent were to foreign companies; of those assigned to U.S. interests, 25 percent went to universities.

Meanwhile, what ever happened to the horror stories that patenting microbes were supposed to engender? Where is the "gruesome parade of horribles" brought in fancy before the Burger Court, or the prospect of the wholly owned cloneman?

One of Collins' co-authors noted sharply in the 1980 Chakrabarty brief that the specter of patent-induced slavery "is quickly laid to rest by reference to the Thirteenth Amendment" forbidding slavery. He went on to point out, "Valid property rights in living entities have been recognized as long as humans have existed, from domesticated goats and plots of Indian corn to today's vast herds of sheep, cattle, and pigs" as well as "the prize bull, whose owner by virtue of a 'monopoly' and current technology earns a good profit while providing a dairy farmer with an opportunity to improve his herd."

NOTES

[1]The American Patent Law Association's *amicus curiae* brief is filed in the Supreme Court of the United States, October term, 1979, No. 79–136, in the case of *Sidney A. Diamond, Commissioner of Patents and Trademarks* v. *Ananda M. Chakrabarty*. American Patent Law Association, 2001 Jefferson Davis Highway, Arlington, VA 22202. Reprinted with permission.

[2]David Masters, *Miracle Drug: The Inner History of Penicillin* (London: Ayer and Spotswood, 1946).

[3]*Chemical & Engineering News,* 1984 article on biotech patents, etc.

7 ACADEME, INC.

AS THE GENETIC REVOLUTION sweeps through the biology and chemistry departments of America's leading research universities, other forces are as radically altering the relations between academia, government, and business. Several changes in federal law and policy have made it possible for private firms to invest in university research to an unheard-of degree, just as the government is sharply curtailing its own research sponsorship. This opening of one door as another closes is difficult to synchronize, and as always major changes in the law raise new questions of ethics and propriety.

The problem here and in other such agreements is not a simple one. Already there have been major concerns that federally sponsored research projects may be "contaminated" with private money, a mix that might lead a private firm to attempt to lock up publicly sponsored research.

Senator Albert Gore has said, "The public policy questions we'll be confronting as a result of the advances in biotechnology are going to be more difficult than any public policy questions we have ever faced. We don't know where the advances are taking us, any more than Prince Henry the Navigator knew where technology would lead his ships."

While in the House of Representatives, Gore headed the Subcommittee on Science and Technology, where he heard and stimulated the debate on how biotechnology ought to proceed. "This map of life itself will probably lead to completely new worlds of which we now have no knowledge," he said. "I think all the participants in this debate have that sense."

What is happening here, he says, is that "the traditional lines of separation between the institution we call the university and the institution we call a private corporation are blending together in a way that has consequences we haven't fully thought through. Not only are they splicing genes—they're splicing institutions and coming up with hybrid forms that raise very serious questions."

To a large extent, the emergence of biotechnology just as this shift is occurring is pure coincidence. But for better or for worse, a great many different events did come together to produce new alliances between the university and the genetic engineering industry. The origin of this upheaval really lies in the conservative political drive to transfer initiative from government to private enterprise. The major expressions of this drive in academia are massive cuts in the National Science Foundation and National Institutes of Health grants that have nourished American research since World War II, passage of the Universities and Small Business Act in 1980, and a change in the federal stance on institutional patent agreements.

According to recent estimates, 20 percent of the funding for what might be called biotechnology research at American universities comes from industry; virtually none did a decade ago. The American Association for the Advancement of Science, to which nearly all U.S. scientists and those with scientific interests belong, recently polled faculty members concerning their ties to industry, their record of publications, and whether or not they kept findings secret as part of contract agreements. The results, published in the AAAS journal *Science,* were full of surprises. First, they showed that faculty members under contract with the new biotech firms were far more productive and more likely to publish than colleagues without such associations. These faculty members, however, were also four times as likely to be *unable* to publish certain research because it involved trade secrecy with a firm.

It isn't necessary to delve into the intricacies of federal policy to understand something of what is happening. We only need look at the world before and after 1980.

Before—Since World War II, most university research has been carried out with federal money. The aim of the federal funding was partly to increase human knowledge, but an equally strong motive was to spur American industry, by making new discoveries that could be transferred to the marketplace. Thus, the government's aim has never been to make money directly on research. Before 1980 the government held nearly all patents granted for work done with public money. Further, if a private company contracted with a university lab that had *any* public money supporting it, the results of the work would almost always go to the government. The aim here was to prevent exclusivity, because the government offered licenses freely on patents it held. The operative philosophy: that U.S. taxpayers should not pay for the same work twice. They already paid for the research through funding.

After—Universities now hold the patents to work done on federal contracts. The operative philosophy: Universities are better able to bridge the gap between laboratory and marketplace, to accomplish the transfer of technology that is still a major aim of government funding. Even more significantly, firms now may contract with universities for exclusive rights to the results of their research, even if some federal money is involved in a given laboratory. The new requirements ban such agreements only when there is significant federal participation in a particular research project. Thus, although universities such as Stanford do not expect to make a lot of money on patent licenses per se, the changes allow for major corporate participation in research—and that does mean a lot of money. And a lot of headaches.

Is the taxpayer paying twice? Not at all, say those who support the new way, because the heavy cost in making new discoveries into reality is in development, not research.

Walter Gilbert says, "Interferon is the perfect example. The basic research in interferon biology consumed a fraction of a million dollars. The applied research required to take it through the stage of clinical testing and all sorts of other steps will take $20 million to $50 million. That's typical."

The taxpayer, in other words, never did pay the development costs, and no company *will* pay them without a guarantee of exclusive rights to the results. That, remember, was a key concern of British drug companies as Howard Florey sought to develop penicillin.

As mentioned earlier, Harvard considered forming its own genetic engineering company. Under the plan, after forming the company the university would lend its name and reputation to the firm and remain a minority shareholder. Whether this might have been a reflection of a great new hybrid or not, the faculty protested that the university had no business in business, and the plan was dropped. Despite the failed attempt, and still urging extreme caution, Harvard President Derek Bok is optimistic about the future of corporate–university liaisons.

Bok told us, "If the United States is to remain competitive in world markets and to prosper economically, we need to improve the capacity of universities to work with corporations to speed the translation of new scientific knowledge into useful goods and services." The universities' obligation to the taxpayer to do so, Bok acknowledges, derives from the "vast public sums" made available for pure research.

Such optimism is not universal in academia, but Bok also sounds a cautionary note concerning the kind of secrecy suggested by the *Science* survey. He notes that "universities need to be extremely vigilant in avoiding relationships with business that will impair the atmosphere needed to carry out first-rate academic research.

"In particular, universities must be careful to forbid restrictions on their freedom to publicize results of research funded by industry, to resist excessive limitations on the right of academic scientists to collaborate freely, and to avoid forms of corporate involvement or corporate ties on the part of their scientists that will induce them to deflect too much of their energies away from basic research to more commercially profitable activities."

Stanford President Donald Kennedy,[1] a biologist, told a congressional committee a few years ago that America's universities with strong research departments have developed in a partnership with government, accomplishing over two-thirds of the basic research done in the country. But he warned that "the enterprise we have built is a fragile one." He urged closer ties with industry, but warned that risks to the free exchange of ideas should not be ignored.

"There is the prospect of significant contamination of the university's basic research enterprise by the introduction of strong commercial motivations," he said, "and potential conflicts of interest on the part of faculty members with respect to their obligations to the corporations in which they have consultancies or equity and their obligations to the university."

Gore is concerned about this possible "contamination" as it relates to universities' relations with Congress. "The public policymaker has often turned to the academic community for advice on uses of institutional agreements. I don't look forward now to sorting through disclosure statements about stock ownership and managerial positions while trying to evaluate the guidance offered."

Nevertheless, Bok and Kennedy see the three-way interchange between the university, government, and business as a major dynamic in the economy and a major requirement for what Kennedy calls "the reindustrialization of high technology." Syndicated columnist Joseph Kraft has written that high technology is the key to restoring the economies of the developed nations: "New products create new jobs, and to a larger extent than commonly realized."[2]

The Brain Drain

The torturous self-examination over the commercialization of genetic engineering hardly marks the first time academics have quarreled with "the hands that feed them." Universities began as creations of churches or royal governments, still later of wealthy industrialists and state and city governments. At each stage throughout history these academies have fought with their benefactors for the right to remain honest and free in the pursuit of knowledge. Thus, if Stanford is now questioning its proper relationship to corporations, it must be remembered that Leland Stanford provided its founding money out of his railroad empire. Does that make all this much ado about nothing? Not according to knowledgeable scientists in both academia and business.

Shing Chang, himself a senior scientist with Cetus and one of the country's leading *B. subtilis* experts, fears the shift to big business may seriously damage the country's research effort if care is not taken. He cites a tragic precedent: German theoretical biology was diverted from its course by industrial chemical interests at a time when it was the best in the world, after World War I. It never recovered. "If we had done that to biology twenty years ago, we'd never have found a restriction enzyme. Nobody would have supported commercial research into something that would seem to have no immediate commercial value."

If there is a danger of research being diverted from its proper aims, there is also a danger of *researchers* abandoning basic for applied investigation. For the first time, Chang noted, candidates for openings at Cetus have the credentials that would have guaranteed good university positions. The big surprise: "It isn't just that they came here instead that marks such a change in trends," he said. "They never even applied to universities for jobs. They just never bothered."

But are these academic researchers being diverted from their original goals by the lure of industrial sponsorship? The *Science* survey suggests that may have happened to some degree—but to a serious degree? Recall David Baltimore's comments concerning the very divergent work done by Collaborative Research, the company he founded and on whose board he sits, and in his M.I.T. laboratory, where he is absorbed in the intricacies of unseeable cancer viruses. That demonstrates that two entirely different kinds of activity are possible and *can be* separated, but many scientists are convinced that often they *are* not.

Dr. Russell Doolittle, chairman of the Chemistry Department at UC San Diego, is highly critical of the trends he sees. Asked if the direction of research is being altered by the flood of "new" money, he said, "I haven't got the faintest doubt. There are tremendous examples: my pristine *E. coli* friends suddenly became interested in blood-clotting when cloning came up—because they could begin cloning factors that there'd be a market for. It's extraordinary."

Doolittle, who "on principle" only consults for the government and nonprofit organizations, says, "I think that the university is the last vestige of a place where people ought to try to maintain some objectivity, where their choice of problems and how they go about solving them is not dictated by financial considerations."

In contrast, he has watched a steady erosion of proper goals: "An awful lot [of effort] is going to be put into making money, and that always means concentrating on the short-range. Nobody's taking the long-range view. I see an awful lot of gloom down the road after a quick fanfare."

Friends tell Doolittle that federal money also has "strings attached," requiring concentration on specific problems. But "the main difference is that the NIH, NSF, and other agencies had not been working on profitability. Their vested interests are much more genuine than playing for bucks." Even so, he said, the government never has supported enough basic research. "There's a lot given to the medical world that I feel is ill-spent. Medical research is the flimsiest kind . . . high school research on a national level."

Instead of rushing after money to replace federal grants, he feels, researchers ought to learn to live with less. "It might make people do a little more thinking. The English have done great work on very little money. They sit around and talk more and drink more tea. But they've done very well."

But few are as sanguine of the prospects for continued high-level research with reduced spending power caused by inflation or—an even worse prospect—reduced total dollars. As we'll see later, others equally knowledgeable in the field believe that irreversible damage to the world's best basic research system is being done through federal cutbacks—damage, like that of a carcinogen, that will show itself many years down the road when it may be too late to repair.

Tom Poulos, while a researcher at UC San Diego five years ago, noted that while much corporate money technically does not require specific projects, "the tendency is to want to do something that the people who gave you the money would find useful." And the recent *Science* survey confirmed that industry ties do influence the choice of research projects, according to respondents.

Further, the sudden prospects of great wealth have brought out the worst in some scientists, Poulos said. "There are some real monsters around. I can't say whether they were nice guys who changed when fabulous wealth came their way, or whether they were always like that but it never showed. But I guess they're like the robber barons. Twenty or thirty years from now we'll say, 'I knew him. He created the billion-dollar ABC company. But he wasn't the kind of guy you'd want to know.' "

Even Stanford's Kennedy, while urging progress in the new relations, points to unwanted side effects. Referring to competition in science, he said in congressional testimony, "I cannot believe that the new terms of trade will not make it much, much worse. Whatever temptations to secrecy and competitiveness are now put in our way by human frailty can, it seems to me, only be amplified by adding the joint temptations of power and profit."

Stephen Atkinson sees a great deal of the research effort at Harvard as executive secretary of the Committee on Patents, and his fear is that few university administrators have any idea of the magnitude of change coming to their campuses. "This and other universities are asleep on their feet about how this is going to affect them," he said. "A lot of positive things are going to happen, yes, but my point is that they aren't aware of the change coming at all."

Evidence of that is the belief that the genetic engineering upheaval is not different from commercialization in the other sciences, Atkinson said. "The Chemistry Department has been consulting for fifteen years. Fine, that means ten or fifteen days a year for very distinct, discrete time periods, and they get paid well. But look at the difference: all these [biology] guys are getting *stock*. The degree of involvement has never been so intimate. You've got a much more intense involvement of these professors in companies, and that's going to make a substantial difference."

Referring to a Harvard professor known to be secretive and aloof, Atkinson said, "The problem isn't that he isn't talking to anybody anymore. The problem is that there are twenty people in his lab who have made agreements not to talk to anybody. How does that affect students? What if they're onto good ideas? Before, if you stole a student's idea, that was plagiarism. Now they grab him and form a company around him."

The *Science* survey confirms all these concerns as legitimate, although the results were on the whole very optimistic.

A New Deal?

There is another view of the sudden demand in genetic engineering: some say it simply represents a welcome shift from a talent-buyer's to a talent-seller's market. Many scientists—including those concerned about the new corporate influences—point out that for years Harvard and other top institutions had a monopoly on the nation's best young scientists. Harvard, for example, retained them for up to seven years, then turned them loose without tenure. The system worked in its time, because with Harvard credentials, those turned loose could almost always find posts at other good universities.

Now, many say, most of the best scientists still head for universities, some of the best look elsewhere, and other job-seekers still have a choice of good jobs. Princeton University's Jacques Fresco and M.I.T.'s David Baltimore say their top students still head into academe, at least for a time. While commenting that it has become "even fashionable" for people to take high-ranking jobs with some of the giant firms like DuPont and Monsanto, Fresco said these are usually senior scientists, not recent graduates. Meanwhile the startup companies, by offering consultancies, are making it easier for academic scientists to earn total incomes closer to what they'd earn if they took industrial positions.

"The salary structure for molecular biologists at good universities already is substantially above the mean for university pay," Fresco said. "I don't think salary has too often driven people to take positions in companies, except when they felt they were being offered significant equity, a sum that would make them wealthy. People with a commitment to their academic pursuits normally still stay with them."

John Abelson, a Cal Tech professor and a principal in the Agouron Institute, an rDNA firm, said the commercial opportunities came at a welcome moment. After the tremendous increases in enrollment of the 1960s and 1970s, university growth has dropped off sharply in the 1980s. Abelson has seen large numbers of bright young graduates suddenly head for DuPont and Monsanto, but these are the "overflow"—good students who would have been squeezed out by the dearth of available faculty jobs. "They want to be assured of facilities for research, and they're getting good projects to work on, too."

Abelson also believes the small rDNA companies are the place where practical, soon-to-be profitable problems should be worked on, the lab the place for more fundamental research. But he does not see a shift in the lab work. He says that, by and large, the top scientists have not left academe to join companies. He offers this reason:

"To leave research now would be the largest of mistakes. If you're not training graduate students, you are not at the forefront of the field, you've lost the fountainhead that will be the source of *future* industrial importance. Grad students talk to their peers, they are excited and imaginative. They carry their professors kicking and screaming into the next generation."

Gilbert saw a split coming in biology that may make many of these concerns passé. "I still do very basic research in my lab at Harvard, but part of the *field* is moving away from that basic level, and that's the work the company does. Molecular biology on the whole I view as splitting. An applied field, including the whole of medicine, will go into the companies. The part of molecular biology stressed in universities is going to be much more about the underlying controls and development of life. The *structure* underlying evolution is now getting more and more discussion—how genetic structure influences evolution."

Moreover, Gilbert believes that attempts to stereotype researchers for pure or applied science are misguided. "One of the reasons you do basic research is that you are self-motivated to find long-range solutions to the problems of society. Now you suddenly come to the point where you can say, well, it doesn't *have* to be long-range, I could actually work directly on something that has a direct application to society. You may, in fact, wish to do that, and it would be part of that same overall view of the world. Is it more worthwhile for me to spend my time applying new techniques to problems that will be useful now, or to work on basic research that will have an impact in twenty years? There's not an automatically correct way of making that judgment."

At Cetus, Shing Chang suggests the form biotechnology might take as the field splits, relating it to engineering, his first academic interest. "I'm partly hoping and partly predicting that an interchange will occur, there will be a turnover in both universities and industry, there will be a high demand for people skilled in basic or applied research. Therefore, if a university wants to train its graduate students for *both,* it should have teachers trained in industry as well as basic research. There doesn't have to be a formal exchange, it may happen as it does in engineering; people quit high-paying industry jobs to teach for a period."

But regardless of method, Chang is concerned that some means be found to support the most basic research. He publishes his findings, as does Genentech, he noted, but by and large scientists in the industry do not. Also, "We do as much basic research here as we can afford to," but most still must be done in universities.

And at Princeton, Jacques Fresco notes that support of that research effort has been waning. So far the problem, though serious, has owed to a lack of *increase* in funding, surprising enough given the academic track record: "The reward for dollars invested by the U.S. government in biomedicine is very high—it even helped to create an industry," Fresco says, and that is one of the aims of federal investment. "But funds over the last fifteen years, especially during heavy inflation, have not kept up. The cost of instruments and supplies have gone up far more than the general cost of living. That adds up to a real *decrease* in funding. The government has not been supporting biomedicine and bioengineering in the way that Japan has."

But the problem could go from serious to acute, Fresco says, because "Gramm-Rudman cuts may result in loss of *absolute* dollar amounts, which would be far worse than any decline relative to 1970 dollars." He adds that already, "The standards for paying NIH research grants have risen to the point where many first-rate projects are not being funded. But consider: Where have the highest proportion of great discoveries in our field been made? In the U.S., because the support for basic research here has generally been better than in Europe."

In November 1986 the National Institutes of Health announced a welcome 20 percent increase in funding for the coming year, and that offered basic researchers a bit of breathing room. But Fresco noted that even that increase—far larger than the administration had requested of Congress—followed several flat years and still will not keep up with the increasing costs of running a scientific laboratory.

Cetus founder Ronald Cape's vision is even more apocalyptic, though he says his and others' warnings before numerous congressional committees have gone unheeded. "The quality of our academic science continues to be the envy of the world," Cape says, and that is purely a function of America's basic research effort. Further, he notes that Cetus and other leading biotech companies "need no help from anyone—it's not about us or about now that we're so worried."

Cape believes that ten to twenty years down the road the United States will lack knowledge that should be developing in basic research today, and he points to an interesting historical case: concern about contamination of the blood supply with AIDS virus, initially a wholly unknown antigen that therefore could not be tested for. That virus is now well known, positive tests have been developed, and a cure may soon be discovered. "None of that could possibly happen had it not been for David Baltimore's very basic research ten to twenty years ago," into the nature of a then-unknown category of viruses that would turn out to include the AIDS virus.

There are two very different reasons for Cape's concern. Foremost, he believes that federal-funding cutbacks to pure, basic university research could prove fatal. "Pure research is the goose that laid the golden egg, and we're killing it," he said. "One of our biggest challenges is to keep the lead in molecular biology for as long as we can." The biological research of twenty years ago has made today's biotechnology industry possible. Today's biological research will determine U.S. leadership or loss of it at the turn of the century, Cape notes. What concerns him for the biotechnology industry is the attitude of politicians and the media toward recombinant DNA companies.

"The media and the Washington politicians seem aware of only one scenario, which they keep replaying: biotechnology is an opportunity for industrialists to get rich, and it's *their* job to protect the public from exploitation," Cape says.

"There has been not one word about the farmers whose crops will be improved and protected, about cancer patients whose lives are being saved, about heart patients being treated with improved drugs. We are on our way to eliminating existing chemical scourges, and there has been not a word of praise from environmentalists. We are perceived by politicians as a lightning rod and nothing more."

We examine some of the real, the possible, and the imagined risks in this new technology in the next chapter. In the long run, we believe, like Cape, that the impact of the genetic engineering industry on university research will be less dramatic and more benign than critics fear. While not dismissing the real possibility that research values may be skewed by the prospect of corporate funding, we believe the more serious threat to America's edge in biotechnology lies in the slow starvation of basic academic research, source of the golden eggs, by reduced funding for experiments and, equally vital, fellowships for training young researchers.

NOTES

[1]Dr. Kennedy's congressional testimony was given to the U.S. House of Representatives Committee on Science and Technology Subcommittee on Investigations and Oversight, June 8, 1981. Kennedy, who provided the transcript, represented Stanford, the Association of American Universities, and the National Association of State Universities and Land-Grant Colleges.

[2]Joseph Kraft, Field Newspapers, 1981.

DANGEROUS ACQUAINTANCES

BEHOLD a popular nightmare:

A research scientist inserts the genes that produce the deadly botulin toxin into a colony of *E. coli.* A few bacteria accept the gene, so they now produce the poison of botulism. Most of them, for one reason or another, are enfeebled by the new gene. One remains virulent and healthy, multiplying in its usual fashion. There is an accident—a dropped flask? an accidental brushing of the colony with a coat sleeve?—and some of the deadly *E. coli* infect the researcher, reproducing in his intestine, as *E. coli* already can do, this time killing him with botulism. Unfortunately, the deadly *E. coli* are very good survivors; they spread through the city, over the countryside, without natural or artificial enemy or antidote, from person to person, in an epidemic unmatched in history.

The authors of this nightmare were not Hollywood screenwriters, but the leading biological scientists in the United States, convening at the Asilomar Conference Center in Pacific Palisades, California, more than ten years ago. It was they who raised the specter of yet another scientific wonder misused to the detriment of the humankind it was to benefit, and it was scientists who derived from this "basic" nightmare what we might call the four categories of risk offered by recombinant DNA, each with its own worst case:

- *Accident*—The opening scenario involved release of a known pathogen, thought to be contained but accidentally spread to the public. Alternatively, we might envision a genetically engineered microbe thought to be harmless, which turns out to be lethal to humans.

- *Environmental Destruction*—A recombinant microbe proves perfectly safe in the laboratory. It is released on an experimental farm where it is supposed to kill insects or weeds, help fix nitrogen, gather essential nutrients from the soil, degrade herbicides or other dangerous chemicals. The microbe is too successful. It knocks out its own natural competitors, destroys beneficial insects or animals, or kills commercially worthless "weeds" that are vital components of the mantle holding soil to earth. The potential for damage here is put most eloquently by Russell Doolittle, chairman of the Department of Chemistry at UC San Diego, whose apocalyptic vision is of a microbe that does precisely what it is supposed to, but "leaves a dust bowl where a farm state used to be."
- *Human Damage*—Eugenics revisited. We begin to map out a Super Race, making decisions that to the believer are God's and even to the atheist ought to result from natural evolution. We monsterize the human race, unable to foresee, in the few short years it will take to accomplish, the irrevocable havoc we are wreaking on our own destiny. In a less extreme but more likely experiment, we tinker with growth genes so that our children will be tall, but without proper regulation this extra growth leads to enormous increases in cancer.
- *Biological Warfare*—No mistakes here, unfortunately. We already have created a nuclear arsenal capable of destroying hundreds of millions of lives in an instant, and every nation that has built such an arsenal has done so in the name of self-defense. No leap of logic is required to argue for developing an arsenal of deadly new pathogens. These would be released only against an invading army, for example, and our own population would be inoculated with secretly devised recombinant vaccine. This is hardly a new idea, but biologicals have always been considered too unpredictable for use in warfare. Nearly all the creations of genetic engineering share a major virtue—they behave more predictably than their natural counterparts. We will cite that virtue frequently in discussing the first two categories of hazards. The two worst features of biological weapons, of course, are that they would be predictable, and any government creating them would likely do so in secret.

Accidents

The two Asilomar conferences in the mid-1970s led directly to the creation of the NIH's Recombinant Advisory Committee (RAC) and a series of guidelines aimed at eliminating risks of the "accident" category. The scientists approached potential problems from two sides: which accidents were likely to occur, and which were of very low probability but of devastating consequence if they should occur; the latter are the so-called low probability occurrence/high risks.

One's fears are not based purely on likelihood of an occurrence, but equally on the gravity of the results if it *should* occur. Thus, the early certainty that there was little chance of a recombinant DNA plague was not reassuring, since the perceived "slim chance" events would be so horrible. The scientists agreed to observe a moratorium on rDNA experiments until the National Institutes of Health—the funding arm for most of their experiments—could set up guidelines. The most hazardous experiments could be prohibited, those adjudged safe could be conducted with minimum regulation—a simple exhortation to use "good laboratory practices"—and various levels of care could be required for those in between.

The guidelines that followed have been in effect since 1976, and although the controversy in the scientific community subsequently died down, the debate is one that reasserts itself periodically. The NIH guidelines require laboratories to operate at "containment levels" coded from P1, for the least risky experiments, to P4, for those potentially most hazardous. The containment levels, which existed previously, now were simply invoked for genetic engineering experiments.

In the summer of 1980, the guidelines were relaxed considerably, and they have been relaxed even further since then. Dr. William Gartland, executive secretary to the Recombinant Advisory Committee (RAC) in the NIH since its formation, explained that the categories have not been altered, but vastly more experiments now may be conducted at lower containment levels.

Most experiments with *E. coli* have been deemed safe, and are thus exempted from these guidelines. Gartland says that means that "the lion's share of DNA work is already technically exempt from the guidelines."

Why? First, the worst case scenario makes several assumptions, notably that an *E. coli* strain, weakened by having to make foreign proteins of any kind, would be able to compete "in the wild" with strains that have evolved over millennia to become adapted to their environments. Further, it assumes that a single colony of such a strain would not merely find an ecological niche—that is, survive as a deadly organism—but would succeed *better than* natural competitors. Since the new *E. coli* would kill its host and thus, ultimately, itself, that does not appear logical.

But this particular point should not cloud the issue: many other scientists who have followed the debate since the beginning do not believe enough attention has been paid to a variety of risks that go beyond the simple if terrifying accident scenario, nor do they feel scientists have made a serious enough attempt to answer long-term questions about the effects of the technology.

The worst case scenario painted earlier should be as familiar as it is frightening, for it has been drawn and redrawn since the nineteenth century, when the achievements of science became the stuff of fiction and prophecy. When Mary Shelley wrote *Frankenstein,* scientists *were* making muscles contract with electric shock, and they *were* watching corpses' hands open and close. But the Frankenstein monster escaped from the lab and into our culture only in fiction.

Similarly, long before Michael Crichton wrote *The Andromeda Strain,* science fiction writers had harvested stories based on new diseases and monster mutants, although until now their favorite prime mover for the disaster has been radiation-caused mutation. Nevertheless, the reality of biological research is in some ways more surprising than fiction.

There are no regulations, federal or otherwise, governing experiments with even the deadliest microorganism, such as those that cause Lassa fever, bubonic plague, or typhus. Genetic engineerings' friends and foes alike ought to pause over that fact. The NIH guidelines, such as they are, apply only to experiments with recombinant DNA, not the natural kinds of bacteria that cause plagues. The NIH's Gartland pointed out that although the government produces a manual of suggested measures for dealing with various pathogens, there are no laws on the subject. There are similarly no federal laws dealing with carcinogens. Containers of hazardous substances shipped interstate must be so labeled. "But you can go home and work in your kitchen on any kind of deadly toxin, and you're not breaking the law," Gartland said. "There is a certain irony that you have this whole regulatory scheme over a theoretical hazard, and no regulations over proven hazards."

Over the past century, laboratory workers *have* occasionally fallen victim to the pathogens they were studying and some have died from their infections, but not once, despite the total lack of regulation, has a laboratory infection spread to the public.

Little news value has been attributed to such laboratory infections, yet they are grim enough. In 1976, Robert M. Pike[1] of the University of Texas Southwestern Medical School, Dallas, published a survey entitled "Laboratory-Associated Infections: Summary and Analysis of 3,921 Cases." Pike gathered information from published reports and form questionnaires sent to researchers as early as 1950. Of his nearly 4,000 cases beginning in 1915, 2,465 occurred in the United States, and 164 of the infections were fatal. A later update by Pike added 158 cases, six of them fatal.

The diseases were as exotic as any imaginable: "Q fever," so named because it raised so many questions among researchers, a disease that is seldom fatal but that can lead to pneumonia; Venezuelan equine encephalitis, tularemia, brucellosis, and psittacosis; hepatitis and typhus; Marburg fever, which mysteriously felled laboratory workers in Marburg, Germany, in 1967, killing seven of them. There is "B virus," which Pike said, "has caused more deaths among laboratory workers than any other virus and it is second to the typhoid bacillus in fatalities due to all agents." Lassa fever, arising from a virus endemic to West Africa and causing inflammation of the organs, "appears to be one of the most hazardous viruses in the laboratory, proving fatal in two out of three laboratory-acquainted infections."

Some have occurred despite precautions, others through carelessness. H. T. Ricketts, whose work in nutritional ailments lent his name to one of them, died in Mexico while studying typhus after lice escaped from an envelope in his pocket, on the way to the laboratory.

In 1970, a woman became infected during an autopsy on a Lassa fever victim in Nigeria. "When the newly discovered virus was brought to the U.S. for study, a female laboratory technician in New Haven, Connecticut, working with the virus was fatally infected. Another worker in the same laboratory suffered a severe infection but recovered," Pike said.

Q fever killed one researcher during a massive outbreak at the NIH's own facility in which 153 workers were stricken, and eight years later a second outbreak infected forty-seven people. In that last episode, six employees of a commercial laundry patronized by one of the workers became ill, and they remain the only "outsiders" known to have fallen victim to a laboratory-initiated disease outside the immediate families of a few researchers. Such an accident would be highly unlikely today because of stricter observance of rules governing lab clothing. Pike pointed out that nearly all the infections occurred before the use of ventilated lab hoods and before sterilization of lab clothing and mandatory showering, all now routine.

At about the same time, Dr. A. G. Wedum[2] of the NIH did a similar survey that was to have major implications for rDNA research; he studied the history of infection at Fort Detrick, Maryland, the army's biological warfare center. Also published in 1976, the study evaluated the chances of an rDNA accident if Fort Detrick's maximum-safety equipment were in use. That equipment would become the basis for defining the P4 maximum containment level in the NIH guidelines.

In various P3 and P4 labs at Fort Detrick between 1959 and 1969, Wedum found one or no infections, and those that did occur were outside the labs' protective devices. One infected worker had been smoking while working at an open benchtop, another brushed his own infected clothing after changing, and still another orally pipetted an antigen and accidentally swallowed some. None died. (While the study indicates that the protective devices did not fail, it does not explain why these workers were not using them.)

Physical protections in the higher containment levels include protective clothing such as gloves, ventilated hoods that draw off airborne pathogens, fully enclosed glove boxes that prevent any contact with the pathogens whatsoever, and negative air pressure in the lab—that is, a slight vacuum—to prevent currents from wafting out. Coupled with this, rDNA researchers now are using enfeebled strains of *E. coli,* such as one named K_{12}, deliberately developed so that it cannot survive outside the laboratory.

Further, many scientists believe that it would be extremely

difficult for them to *deliberately* create a recombinant organism that would stand a chance of competing in the wild, and that even the deadliest of recombinants designed with protection in mind would die outside its petri-dish culture. Given the precautions, they rule out the kind of widespread calamity once envisioned. One scientist predicted, "We'll someday need our P3 lab—to protect our strains from being killed off by a hostile outside world that might get in here, rather than the other way around."

Bacteria mutate all the time naturally, but so few mutants survive that, in nature, mutation is generally synonymous with death. On the rarest of occasions, a natural mutant is suitably enough adapted to its environment that it can win its own niche among all the strains already competing there. Artificial recombinants would stand still less chance of surviving.

Sheldon Krimsky, an early non-science RAC member and a critic of uncontrolled genetic engineering, agreed that the likelihood of a disastrous accident is small and that "history shows that it is not the laboratory which produces the greatest damage." Krimsky is a professor of urban and environmental policy at Tufts University in Massachusetts. He noted that in terms of biohazards the governmental "institutions developed to look into them were formidable, fair, and reasonable, even though I didn't agree with all the decisions made within those institutions." His major complaint is that, while everyone talks of the unlikelihood of disaster, "I see no efforts made to see if the worst case scenario would work—putting the genes for botulin into *E. coli,* and really trying to see if under extraordinarily controlled circumstances, you *could* get it to survive and multiply. They say they couldn't, but when they begin to do experiments like that and do them honestly, I'll begin to be convinced that this stuff is not less hazardous than botulism all by itself. But those experiments have not been done. They're fearful that if they do them, they'll find out too much."

That uneasiness extends to others who generally applaud the work of genetic engineering and its promise for the future. Bruce Hilton, director of the National Center for Bioethics in Richmond, California, said that despite reassurances of safety, "it still gnaws at the back of my mind: have those fears really been resolved, or is this a drawing together of members of the science community, fearing that they may confront another Galileo- or Copernicus-situation unless they pull together." His uneasiness is heightened by sensing "terrific pressure from the pharmaceutical companies and labs to get on with it and not be tied down with restrictive research rules." In any event, he says, "They seem to relax the rules more every six months."

Gartland, an executor rather than formulator of policy, is cautious in assessing the changes and proposals. But he notes that making compliance voluntary "in effect marks a dismantling of the whole system, and that has some people worried. Because if you dismantle it and something happens, then we are back to ground zero again: 1974. You've got no national committee and no national forum for anything, and that's a very dangerous thing."

Perhaps most importantly, the guidelines now apply only to work done under federal contract. As Gartland puts it, "All we can do is 'follow the federal dollar,' to say that following the guidelines is a condition for your getting this money." As more and more recombinant DNA work goes into uncontrolled industry, the limitations of such regulations make less sense. Industries have agreed to follow the guidelines, but that already is voluntary, except in a few cities that have ordinances governing genetic engineering. Gartland pointed out in late 1981 that a multitude of congressional efforts to pass national legislation regulating recombinant DNA work has gotten nowhere. "It was decided that no federal agency had complete jurisdiction—the Occupational Safety and Health Administration comes closest, because it can require that you provide a safe workplace." But even OSHA has no specific charge with regard to laboratory safety. "Therefore, national legislation was needed. But no law has ever been passed. It remains in limbo to this day." That has changed now—but with no diminution in the controversy over what the rules *ought* to be

Environmental Damage

The storm over genetic engineering reported daily in America's newspapers beginning in about 1984 has been almost exclusively over microbes devised to be released into the environment. But in order to sort the real from illusory risks of such release, we must consider what we learned in contemplating the accident scenario: genetically engineered microbes behave more predictably than their wild counterparts. That, in fact, is the purpose of genetically engineering them. Further, when a microbe is engineered to produce a protein of benefit to us, it is using enormous quantities of energy that it otherwise would be using to reproduce its own proteins—that is, competing for survival.

Bacteria in the wild mutate constantly. Their rapid evolution allows them to adapt to changes in their environment, and as we've pointed out, by exchanging plasmids, bacteria in effect are able to confer on one another semipermanent genetic changes to adapt even better to environmental change—and to thwart our efforts to eradicate those that make us sick.

Biologists have used this mutability of bacteria for decades; "new and improved" antibiotics are developed through a procedure called forced mutation in which a colony is stressed so that it mutates more rapidly, and the colony is grown in an environment favoring the mutant that does what the research is after. Genetic engineering doesn't even necessarily accomplish this faster—when it works, it simply accomplishes the goal more predictably.

Let's consider the risks to the environment in two phases: the long-term potential for genetic engineering to cause environmental damage, and then, importantly, the risk to the environment from those experimental releases already proposed, and which have been the subjects of so much controversy.

UC San Diego's Russell Doolittle argues that, just as organic chemistry and other branches of science rushed too quickly to the marketplace, industrial genetic engineering will be marked by a brief period of bright hope followed by environmental ruin. We won't likely set loose a deadly new bug. "We'll create a dustbowl where a farm state used to be. It won't happen because something crept out of the lab. It'll happen because in our rush, we didn't think of all the consequences of doing something that looked good at the time—something which in the short run will *be* good." Such thoughtless advance is bound to bring disaster, Doolittle feels.

"I've never seen any of these things *not* backfire, just like DDT." Noting that he believes the "nightmare" stories of the risks of recombinant DNA are overblown, he said: "No, the real hazard is that we're going to create a dustbowl doing something that *works*."

For Doolittle, proper precedents to consider would be the pollution of the atmosphere by automobiles, or the terrible erosion caused by slash-and-burn techniques in creating new farmland.

"Who watched that first internal combustion engine roar and thought, 'I wonder if that'll clog the atmosphere with junk, kill plants, foul rivers?' Who could have? I'm afraid that in this case just such mundane failure to look ahead will cause the real disasters."

Sheldon Krimsky also believes that industrial scale-up is precisely the action that will usher in new hazards through "the purposeful dissemination of materials in the environment with inadvertent consequences."

Bruce Hilton asks, if plants that can fix their own nitrogen come to dominate even a branch of agriculture, could they also change the nitrogen balance of the air? If so, what would the consequences be? If that seems farfetched, Hilton notes, who would have believed aerosol-spray cans could affect the ozone layer?

The suggestion made repeatedly by those concerned is not that investigation be halted, but that the long-term implications of genetic engineering *applications* be studied. Krimsky is especially critical of the government in this respect. Of such study, he says, "the one institution that we have in this country designed to do that, the Office of Technology Assessment, has completely relinquished any of its capacity to do so."

He said the OTA's "fundamental interest has been: Can we compete with other countries? and, How can Congress stimulate the development of biotechnology?" Krimsky adds, "That's perfectly legitimate, there's nothing wrong with that, but it has to be balanced with some assessment of how these things are going to alter society. Do we want to undertake a system for producing biological pesticides when we had such an outrageous situation with organic chemicals?"

Organic chemistry, moreover, produced few hazards in the *laboratory*—certainly none that could not be handled. It was as an industry that it left us with disaster along the Love Canal in New York and contaminated municipal wells in the Southwest, not to mention providing those once universally favored aerosols.

It is important to point out that Doolittle, Hilton, and Krimsky made all these comments to us more than five years ago, and they were published in the first edition of *The Gene Age* in 1983. Far from being "shots from the hip," these have long been the legitimate concerns of those attempting to see the future of genetic engineering. And those commenting, far from being alarmists, are, respectively, a biochemist, a Methodist minister, and an environmental scholar. In fact, all that we take exception to is the sense of inevitability that Doolittle expresses, which we will explain.

Organic chemistry has given us several environmental disasters while providing benefits modern society could hardly surrender—and environmental damage continues despite heavy regulation. Experimentation with pathogens and microbial poisons (which is what antibiotics are) has not led to such disasters—and this field has been virtually unregulated except when deliberate human consumption is involved.

A key difference: now we are looking at the deliberate release of a nonnatural microbe into the environment. Can we forecast what it will do? *Never with certainty.* And that means that there always will be a certain risk that bacteria will damage crops in the course of curing a pestilence, and methods of fighting those bacteria will have to be devised. All we can do—all we can ever do when confronting a potential boon—is to make every effort to minimize the risk, and to ask ourselves if the potential benefits outweigh the risks. Certainly, in those cases in which a release of microorganisms might harm people, extra delays ought to be required to make sure that every precaution is taken and that the benefits would be extraordinary. And in those cases in which a harmful side effect of release is expected, release should be forbidden and use of the microbe regulated.

Let's consider some current and near-future examples of proposed release of genetically engineered microbes as examples of potential hazards.

This is the arena of Jeremy Rifkin, the incredibly influential activist, deemed gadfly, obstructionist, hope-for or destroyer-of the future, depending on who is "deeming." Most scientists are appalled by the unsoundness of his scientific attacks of genetic engineering proposals, yet many of them say his criticisms correctly pinpoint procedural flaws that could be hazardous. Most of those with any concern for genetic engineering say he has hijacked every relevant issue, adding a shrill and unreasoned dimension to the sober reflections of people like Doolittle, Hilton, and Krimsky. Yet Krimsky, for one, admires Rifkin's success in getting these key issues in the public eye, which virtually no one else has been able to do. In the following examples, Rifkin has filed about half a dozen lawsuits in a variety of federal courts; there is virtually no environmentally related activity in genetic engineering that Rifkin has not opposed through his Foundation on Economic Trends. We will not note his opposition; it goes without saying. Later, we'll comment on Rifkin's role.

Dioxin-eater Consider the development of a microbe that would devour dioxin as an example of environmental release that ought to be approved—if the microbe is ever successfully created; many scientists are trying. At its worst, dioxin lives in the soil in concentrations of a few parts per million. Thus our magic bug would have to be able to live on more than dioxin or it wouldn't reproduce well enough to survive. Once the soil were depleted of dioxin, it would have no better odds on surviving and reproducing than another, native bacterium. And finally, most importantly, whatever harm might be done by the bacterium would be nil compared to the harm done by living on top of a dioxin wasteland.

Ice-Minus Snowflakes and raindrops form around particles of dust, which are then said to "nucleate" them, that is, provide a nucleus for their formation. Ice particles can form around such dust motes, but they also form around certain bacteria, and that has an unfortunate consequence in agriculture. The bacteria inhabit potato and strawberry plants, without directly harming them; however, at a temperature just above freezing the bacteria nucleate ice out of the water vapor in the air, freeze-killing millions of dollars' worth of the nation's potato and strawberry crops every year.

Enter the genetic engineer. This nucleation potential is caused by the protein product of a single gene in the bacterium, so scientist Steven Lindow engineered a strain of the microorganism by snipping out that gene and sought permission to field-test it on an experimental potato patch, triggering Jeremy Rifkin's first lawsuit in the case, in 1983. A short while later, Advanced Genetic Systems (AGS), a Berkeley, California, rDNA firm with which Lindow initially worked, sought to field-test the microbe on a strawberry patch in a remote area of California in early 1986.

Some critics charged that the EPA should not have approved the field tests because not enough was known of the harmful potential of the new bacterium. For example, what if it were radically more successful than the wild types, either forcing them out of existence or establishing itself powerfully in a niche. Then, what if this new strain proved damaging to other dwellers in that niche? Finally, if the new bacterium were a radical reproductive success, and ice no longer were nucleated out of the water vapor, then couldn't a real change in weather patterns result?

Answer: The "novel" bacterium exists in nature already, one of thousands if not millions of genetic variants of this strain—remembering that bacteria mutate spontaneously, constantly, gaining and shedding new genes almost as regularly as they reproduce. If the ice-minus bacterium were going to establish itself in northern California, in other words, it already would have done so. *In this case,* it would be ludicrous to suggest that a bacterium that has nothing bred into it to improve its survivability could change the weather; it would be reasonable to assume that the strain would die back to some natural level—but hopefully not before it could be determined if, in fact, the experiment worked and ice did not form on the experimental plants.

The EPA approved the field test on this basis, yet the charge continued to be leveled that the environment was at risk, local opposition remained steadfast, even increased and spread, and the experiment never occurred. This is unfortunate, because ice-minus offers the perfect case for monitoring rather than prohibiting by regulation as the means to minimize risk. Regulations, in other words, already were sufficient to ensure that *this* experiment would not harm the environment, the potential benefit to society in developing such a successful strain could be large, and the government monitored the proposal well. Such monitoring ought to continue, and especially cover some *future* experiment in which scientists would have to address the possibility that a strain of the bacterium deliberately engineered to be successful at *living* in its environment (as well as nucleating ice) would not take over its niche with harmful, unpreventable consequences. In other words, a close watch ought to be the preventive medicine of choice, rather than injunction. Yet this and other experiments are being enjoined, meaning that their benefits will never be known.

Ice-Minus II While awaiting approval to field-test ice-minus, AGS introduced its altered strain into some plants atop the roof of its Berkeley laboratories. When the EPA learned of the test, the agency sharply criticized the firm, suspended the by-then-issued field test permit for ice-minus, and levied a fine of $20,000 for violating regulations.

Critics of genetic engineering point to the incident as an inherent hazard in biotechnology: If something harmful had leaked into the environment, no amount of regulation would have prevented it.

That is true. However, the stiff fine and the resounding criticism—industrywide as well as nationwide—ought to serve as proper warning to any other firm contemplating such a move. In fact, Advanced Genetic Systems said the experiment was conducted because the company did not believe the work constituted "environmental release," not because it was trying to circumvent the regulations. Now that has since been established.

The Cutworm Debacle Monsanto has developed a bacterium whose natural version dwells in the soil; the engineered version produces a potent insecticide. Better, the insecticide is harmless to humans and animals and causes no harm to crops. Monsanto applied for EPA permission to field-test this bacterium, which it hoped would be especially useful against destructive cutworms, and the case is still pending. First, one must ask the same questions as of other insecticides: Would the natural toxin kill useful insects as well as pests? Result in the extinction of other animals—fish, for example—that live on insects? And now the question that does not need to be asked of a factory-made poison: Will it spread? Monsanto says the bacterium will die back after the crop is harvested. According to a variety of news reports, the EPA is not convinced of that point, and if so that is a good reason to withhold approval.

The issue, however, appears in limbo if not dead, not for that reason, but because of the storm over the AGS field test and the general climate against any environmental release of anything, fostered by some environmental groups but most intensely by Jeremy Rifkin. In other words, strident opposition appears to have overshadowed reasonable considerations, and if so, that is an unfortunate case of throwing the baby out with the bathwater.

Guarded Welcome

Cambridge became the focus of controversy over rDNA experiments almost as soon as the NIH guidelines were published, and with good reason. As home of both Harvard and M.I.T., the city fathers correctly assumed, Cambridge would become home to many rDNA firms that would not have to obey the federal guidelines.

The battle on the city council was characterized by as many wild, heated charges as logical arguments, with then-Mayor Alfred Velucci forecasting that the city would be plagued by monsters and epidemics. For all the fury of the public hearings, however, former Councilman David Wylie recalls that only a handful of people were present on April 27, 1981, when the council passed the nation's first ordinance[3] governing recombinant DNA, and it is a marvel of simplicity.

Basically, the ordinance requires anyone doing genetic engineering research to agree to follow the NIH guidelines and to get approval from an advisory committee of the council before doing recombinant DNA work. Both lay people and scientists make up that committee. In 1981, the ordinance was updated to cover the larger scale rDNA work expected from the cluster of new firms in the city; they must now get licenses and follow newer NIH guidelines governing recombinant experiments in larger quantities.

Macy Koehler, who became Harvard's biology safety officer when the post was created in 1977 and served until 1979, praised the ordinance both because it will protect the public from accidents and because it will protect researchers from misunderstandings about their work. As safety officer for those first years, Koehler made sure that not only the NIH rules but university biosafety rules were followed throughout the Harvard Biological Laboratories.

"I believe in public involvement in scientific decision making," the biologist said just after the ordinance went into effect in the summer of 1981. "We must make an effort to communicate to nonscientists, to tell them what we're doing and why. But nonscientists must make an effort, too. Even when put in the very simplest language possible, the subjects [of genetic engineering] are very difficult."

Boston soon followed Cambridge with an ordinance, and several more cities have adopted them since—but they remain a handful nationwide, and that is ironic given the passion that attends to individual genetic engineering proposals.

On the other hand, the Cambridge story does illustrate how smoothly the first phase of industrial engineering has gone. Dr. Melvin Chafen, city commissioner of health and hospitals, says there have been no problems with fourteen licensees—university and corporate—since the ordinance passed, nor has there been public outcry over any issue. Velucci, still a city council member, remains concerned over the future of genetic engineering, but has not opposed its growth in Cambridge, an industrial city that badly needed new firms.

An important note: the Cambridge ordinance specifically forbids environmental release of microorganisms. The only firm permitted to build a greenhouse—BioTechnica International—had to design it with such stringent safeguards that it is in effect a high-level containment laboratory, to guard against any possibility of escape of a microorganism.

Against this background, let's consider the new regulations.

Put in force in June 1986, new federal policy puts responsibility for genetically engineered organisms under one of these agencies, depending on the use of the product.

Food and Drug Administration: Regulates all pharmaceuticals, as previously had been the case. The FDA is certainly the toughest federal regulatory body, and companies wanting to manufacture any product for human consumption will face the toughest hurdles. It would be hard to imagine opposition to this assignment.

Environmental Protection Agency: Oversees any genetically engineered microorganism destined for release into the environment, plus some others. The EPA will share authority with USDA on microbes intended for agricultural purposes.

U.S. Department of Agriculture: Monitors genetically engineered crop plants and animal vaccines made through recombinant DNA, in addition to microbes aimed at enhancing or protecting crops.

The National Institutes of Health: The Recombinant Advisory Committee of the NIH will continue to have primary responsibility for university research although, as noted, so much basic research is already exempt from federal control that the RAC's work has diminished considerably.

Occupational Safety and Health Administration: Any genetically engineered microbes used in the workplace will come under OSHA's purview.

Criticism of the new regulations has come from two directions. Scientists and those in industry, while generally pleased with the clarification of responsibility, are concerned that in those areas in which agency role is not clear there will continue to be a bureaucratic quagmire over "who's in charge here." Senator Albert Gore has said he believes the regulations have done little to solve that problem.

For example, would a genetically altered pesticide like that proposed by Monsanto come under EPA, USDA, or OSHA regulation, since farms are very definitely workplaces to OSHA? Those in the industry are only concerned here that they know which agency to apply to and that when approval is given, it is final. Otherwise, the industry will stagnate.

On the other hand, the Reagan policy has been criticized sharply by several environmental groups and not a few scientists for being too soft in critical areas, especially those involving gene deletions and so-called "regulator manipulation."

For example, microbes that merely involve deletions of genes are not subject to review, because nothing has been added to the microbe, novel as it may be. This seems a reasonable approach to us. As we noted, microorganisms shed genes regularly in nature, and their only "evolutionary strategy" is to reproduce constantly. It is hard to imagine any gene loss that has not occurred—and any that have not proved fatal will be represented in some subpopulation of the particular bacterium.

Under the new guidelines, experiments with microbes are exempted from review if the genetic tinkering involves only addition or alteration of regulator gene sequences. We've already discussed several of these, including the origin of replication on DNA, and the ribosome binding sites for messenger RNA. Overproduction, the key technique to industrial genetic engineering, involves manipulating these sequences to maximize the expression of the genes they control, and therefore to maximize production of the desired protein. Initially, it seems a gross relaxation of standards to exempt all work involving regulators. Remembering that these are microbial regulators we are discussing, however, the same principle applies in some cases as with gene deletions: If regulation is simply removed, we can assume that we are doing nothing that hasn't happened many, many times over the course of the microorganism's evolution. If it was fatal alteration, as most are, the organism won't survive in the wild; if it was a beneficial one, the organism is already out there.

What we've just said, however, is hardly all-inclusive of the regulator manipulation possible with genetic engineering. What if we tinkered with a microbe's regulator genes in a more complex fashion; for example, by introducing a regulator in a way that enhances the ability of our novel form to reproduce, a way that natural evolution likely would not confer on it? Such experiments ought to be as strictly monitored as any whose specific aim is to enhance a microbe's ability to survive *in the wild.*

A key final point concerning regulation. It is generally easy to engineer a microbe so that it *cannot* survive in the wild. When we discuss engineering microbes that digest pollutants in the final chapter, you will see this strategy at work. Even so, microbes engineered to digest a pollutant whose "deadly" concentrations number a few parts per million have no great survival advantage over natural strains.

This is another example of our thrust toward monitoring rather than trying to anticipate all problems in experiments in order to forbid them in advance. Close monitoring would ensure that microbes perform as advertised, die back as advertised, and carry no unwanted side effects before they are put to widespread use. Overly enthusiastic prior regulation costs us the enormous benefits that genetic engineering can bring.

The regulations as a whole have been attacked and legally contested by several different groups, but generally under the leadership of Jeremy Rifkin and, as noted earlier, no account of the state of regulation in genetic engineering today would be coherent without reference to him. Who is he? First, he is a man of fairly impressive credentials—but none are science related. An economics graduate of the University of Pennsylvania, a master's graduate in international affairs from Tufts University, he is the author of nine books on a variety of subjects but all aimed at philosophical theories of knowledge and the role of science in society.

In a wide-ranging, congenial telephone interview he articulated his opposition to one after another of the genetic engineering proposals against which he has filed lawsuits. Many of Rifkin's arguments we have already spelled out; the difference is that Rifkin has taken legal action against those attempting, in his view, to circumvent proper procedure. Rifkin's many supporters, and even many of those in genetic engineering who oppose him, say he has performed a vital function in forcing both scientists and government to follow proper procedure. And one executive involved in regulation for a major biotech firm noted, "When you call genetic engineering companies concerning regulation, we offer no comment or nothing for the record. But Jeremy Rifkin always has a comment and it's always on the record."

Rifkin says he began the Foundation on Economic Trends a decade ago, when its mission of acting as watchdog over genetic engineering had not emerged as more important than others; hence the name, which he doesn't like but feels stuck with, "since it's so well known." Probably no one in America has such a conflicting image as Rifkin: good-humored and articulate in person, shrill and frequently less than rational in his scientific commentary.

Known to journalists and their audiences as spokesman for "the other side" whenever genetic engineering is discussed, Rifkin does not think of himself as an activist so much as an intellectual force. He guest-lectures about genetic engineering on campuses so frequently that he says he now has visited three hundred universities, or one out of every ten in America. There, he is regarded as an engaging, charismatic speaker. Sheldon Krimsky says he and other audience members were rapt as Rifkin addressed a meeting of biotechnology critics in Washington recently, because of Rifkin's way of engaging the audience, of bringing about participation. Rifkin: "I get bored just lecturing at the podium; the audience does too. So I try to engage them, to make them think." A file he provided us had numerous letters of praise from university educators for the way in which he got his student-audiences engaged in his subjects.

He was every bit as engaging in conversation, and pointed out, correctly as far as we can determine, "Contrary to the way my critics behave, I have never attacked a person, or said that one side or another was good or evil. . . . My objective is to bring out the good and evil in what is happening, in what can happen, not in the people involved." He offers a long comment on the dangers to our human inheritance if we genetically engineer solutions to our problems with the environment into people, rather than attacking the problems themselves. Then he concludes, "But I'm no better than anyone else. I go for the quick fix, too. That's what I'm saying, it's what we all tend to do, that's why it's so dangerous."

True enough. Further, Krimsky believes that Rifkin has modified what was an indiscriminate opposition to all genetic engineering, perhaps has even "mellowed" somewhat; and Rifkin himself says that with time environmental concerns have come to the forefront and his earlier fears concerning such nightmares as enforced, Hitler-style eugenics have abated.

Nevertheless, it is difficult to find in his voluminous writings, his ubiquitous appearances as spokesman for the opposition, or in a long conversation any environmental application of genetic engineering which he does *not* oppose. This is truly unfortunate. We and others have detailed the benefits that can accrue from environmental applications of biotechnology; they are not simply economic, although one can cogently argue that at this stage in the industrialization of the West the economic development alone would warrant a progressive if prudent attitude toward such proposals. Despite his statements to the contrary, all of Rifkin's activities as gadfly appear aimed at bringing genetic engineering to a screeching halt; that is how he is perceived, and the proliferation of court suits in his name attests to this obstructionism even as he argues to the contrary.

In an illuminating conversation in *Science Digest,*[4] the magazine's interviewer put quite succinctly the dual image of Rifkin that emerges:

"When you speak in public, you take a very pragmatic view of biotechnology. You're careful to describe the potential payoffs. You mainly discuss questions of implementation—how we will control what is done or who will decide on it.

"But in your books and legal maneuvers you attack science and call for an outright ban on genetic engineering. Will the real Jeremy Rifkin please stand up?"

Rifkin's reply is that he does not oppose scientific inquiry, but on intellectual grounds opposes the kind of science that seeks to dominate nature, a science alienated from and at odds with the environment. He suggests, "Knowledge is empathy with the environment, empathy allows us to establish a new type of security by becoming a member in good standing in the community of life."

Beautiful. Now listen to the broadside in which he describes himself to potential lecture audiences:

"Jeremy Rifkin is one of the most vociferous and effective opponents of genetic engineering and nuclear technology. In the last few years he has single-handedly forestalled the emergence of an entire scientific revolution. Millions of dollars have been put on hold, scientific experiments have been blocked, and shackles have been placed on new fields of commercial development." This crowing finished, the flyer concludes moderately: "Rifkin challenges our scientific worldview and the high-tech optimism of the 'Age of Progress.' He dares to question many of the most fundamental assumptions of contemporary Western civilization."[5]

Stephen Jay Gould, a Harvard biologist and one of America's finest writers on the life sciences, wrote in a review of Rifkin's book *Algeny:* "Among books promoted as serious intellectual statements by important thinkers, I don't think I've ever read a shoddier work. Damned shame, too, because the deep issue is troubling and I do not disagree with Rifkin's basic plea for respecting the integrity of evolutionary lineages."[6]

We come away from considering many of Rifkin's activities with much the same feeling: *Damned shame, too, because* . . .

The Jeremy Rifkin who led protestors at a National Academy of Sciences meeting by chanting, "We will not be cloned" is the same man who has brought more public attention than anyone to the potential hazards of biological warfare. The Rifkin who alarmed the populace by claiming that a minor experiment with ice-minus in California might alter weather patterns is the Rifkin who has held scientists' feet to the fire on the need for proper procedure. (Changes in local environment on a *large* scale can alter weather patterns. Large urban areas do so, as is well known.) And the Rifkin who claimed that allowing microbes to be patented would lead to patented humans is the same who has done far more to bring the *issues* of genetic engineering to the public's attention than any of the reasoned voices we have quoted here—yet forever raises the issues under a screaming headline of his own device announcing a four-alarm fire.

Among the issues, none holds a candle in importance to the one that follows.

Engineering Evolution

To many scientists and religious leaders, even the fears of environmental damage are dwarfed by the major ethical questions raised by genetic engineering, by such possibilities as our altering the human race significantly to create new kinds of humans. To repeat Shakespeare's phrase with Huxley's construction, "O brave new world that has such people in it." Some of the fears voiced have been hysterical. The U.S. Constitution, as we said earlier, cannot be repealed by scientists. But real pressures can be put on all of us to change that Constitution, and many fear we are by no means prepared to deal with the challenges we face.

Genetic counseling is already reality, ushered in by the ability to detect key genetic factors through amniocentesis and the Supreme Court ruling permitting abortions. Bruce Hilton is such a genetic counselor, a Methodist minister who spent two years at the Hastings Institute during which time he was a senior editor for the book *Ethical Issues in Human Genetics*. He now publishes a newsletter on bioethics as well as directing the center. Hilton refers to this most serious genetic engineering as "short-circuiting evolution."

"What happens when you jump several million years from one organism to the next? Maybe nothing, but nobody in science seems to be addressing questions like these," he says. Hilton says he rarely mentions that he is a minister because "scientists are suspicious of religion's old role" in suppressing inquiry, and he is very much in favor of the progress promised in biology. "I'm in favor of amniocentesis and a woman's right to an abortion," he says. But he sees dangers lurking, some of which already have revealed themselves.

For example, some states have tried to require black couples applying for marriage licenses to be screened for the sickle-cell anemia gene. The couples may marry whether they carry the gene or not, but Hilton sees such forced screening as a possible prelude to something far deadlier—either legal bars to such marriages or forced abortion. He pointed out that 8,000 babies are born in the United States every year with Down's syndrome (mongolism), and it costs about $250,000 to institutionalize most of them over the course of their lives—that is, about $2 billion a year figured over an indefinite period. Half these babies are born to an identifiable group—women over forty years old. Hilton is concerned that eventually there could be pressures to force women identified as carrying Down's syndrome babies to have abortions, to save $1 billion a year of those costs.

"I certainly would not be in favor of any such idea. Society can bear the cost more easily than it can bear such a striking seizure of the freedom to procreate," Hilton said. "But these pressures would not be new." Further, he said, "There are some serious, thoughtful people on the other side."

Charles Frankel wrote in *Commentary* in 1974 that a deliberate intention to create a "new man" unites all revolutions of the right and left.

More pointedly Frankel says, "The partisans of large-scale eugenics planning, the Nazis aside, have usually been people of notable humanitarian sentiments. They seem not to hear themselves. It is that other music they hear, the music that says there shall be nothing random in the world, nothing independent, nothing moved by its own vitality, nothing out of keeping with some idea: even our children must be not our progeny but our creations."[7]

Such arguments are, indeed, all part of a larger view fought over for much of this century concerning "eugenics," and those favoring strict procreation control formed an extremely powerful group that predated Adolph Hitler, but one which is presumed to have given him many of his notions on race and racial purity. Eugenicists have included presidents of major universities, leading liberal intellectuals, industrialists, and political leaders, and they were extremely influential through the 1920s. At the heart of their thinking, Hilton noted, is a belief in human perfection. He added, "It is part of western lay theology that it is possible to cure death and that it is possible to make a perfect human. Both are wrong, but I'm convinced these notions exist in our subconscious."

Hilton's point: the major teachings of eugenicists may have fallen out of favor, but they remain near the surface, and there is reason to be concerned that genetic engineering and other forms of biotechnology will bring them to the fore again unless there is adequate discussion of the issues. Certainly, genetic engineering might detect the crippling diseases Watson referred to at such an early stage that all but those who oppose abortion on purely religious grounds would come to strongly favor it.

Hilton said that R. G. Edwards, a British pioneer in developing the technology for the first "test-tube baby," told him that human embryos can now be "typed" for genetic illnesses and sex when they have split to only eight cells. Further, if it becomes possible to alter the genetic *makeup* of embryos in some early stage of development, for the first time in history genetic illnesses and many birth defects would be curable.

It might seem hard to imagine a test that would detect pregnancy at such an early, hours-old stage as an eight-celled embryo. But if eventually one were available, it would be a boon to women at high risk of bearing defective children.

Five years ago, the prospects for gene therapy seemed remote enough that we could speculate on the moral issues involved. Now, as we have seen, gene therapy is at our doorstep. But there is a key distinction that must be made in the forms of therapy that all derive from genetic engineering. First, there is administration of human-gene products that have been manufactured in bacteria. That has been going on routinely since human insulin went on the market several years ago. We have taken that therapy a questionable step further, however, when we consider giving human growth hormones to children whom we simply want to be advantageously tall.

But the distinction between this administration of gene products and gene therapy is a major one, and there are in turn two forms of gene therapy radically different from one another in their ethical implications. Somatic-cell therapy refers to alterations in one person's genes that would affect *only* that person—for example, correcting a genetic error in somatic genes that places the wrong amino acid in a protein—and the correction goes no farther than one human body. The second form of gene therapy, one surely to be controlled far more rigidly, involves altering genes in someone's germ line cells. These are the cells of inheritance. Genetically engineer in the germ line, and you change the genes someone passes on to his or her progeny. Now you are doctoring not a human life but human destiny.

What is difficult in confronting these decisions, of course, is that they grade into one another. If it is all right to give human growth hormone to a child who would otherwise be a dwarf, then it seems all right to provide that child gene therapy to accomplish the same goal. Then is it all right to engineer the correct combination of genes into that child's germ line, since dwarfism is hereditary? Is it all right to engineer some combination of growth genes into our ordinary children's germ lines, to make sure their progeny are suitably tall? The trouble is, as everyone from Russell Doolittle to Jeremy Rifkin knows too well, we all *do* like the quick fix.

The coming of the Gene Age will not find us wholly without legal and moral guidance on these major issues. Experimentation with human subjects, for example, has long been strictly regulated by the government, and FDA requirements for bringing new drugs to market are estimated to take an average of eight years to fulfill. But neither are the questions idle or premature. We appear to be on the verge of cutting the umbilical cord to nature, freeing ourselves from muscle power and industrial power, perhaps even from the ties that bind us to the natural and slow process of evolution.

Ray Thornton, president of Arkansas State University and an early member of the RAC, observed that people are most comfortable with science "which explains how things work, which promises health, physical well-being, and material progress."[8] But now and then, he reminded, science bursts its bounds and leaps into areas that challenge our concepts of life and the institutions we have built in service of our lives. "When Galileo offered the theory that the Earth revolves around the sun, it was bad enough to his contemporaries that [in their opinion] he committed scientific error. It was worse that he committed heresy as well."

The questions of genetic engineering that most chill our souls, too, involve not error or chance, but the violation of our concepts of life itself—heresies to believers and atheists alike.

NOTES

[1] "Laboratory-Associated Infections: Summary and Analysis of 3,921 Cases," by Robert M. Pike, *Health Laboratory Science,* Vol. 13, No. 2, April 1976, pp. 105–114; also, "Laboratory-Associated Infections: Incidence, Fatalities, Causes, and Prevention," Robert M. Pike, *Annual Review of Microbiology,* 1979.

[2] "The Detrick Experience as a Guide to the Probable Efficacy of P4 Microbiological Containment Facilities for Studies on Microbial Recombinant DNA Molecules," A. G. Wedum, concluding research sponsored by the National Cancer Institute. (Manuscript of January 20, 1976, provided by NIH.)

[3] "Ordinance for the Use of Recombinant DNA Technology in the City of Cambridge," No. 955, passed April 27, 1981.

[4] "Jeremy Rifkin, Devil's Advocate," *Science Digest,* May 1985.

[5] "Jeremy Rifkin/College Lecture Tour," flyer published by the Foundation on Economic Trends (provided by Rifkin).

[6] Gould review of *Algeny* in *Discover,* quoted from *Newsweek,* May 28, 1986.

[7] "The Specter of Eugenics," by Charles Frankel, was reprinted from *Commentary,* March 1974, by permission; all rights reserved.

[8] Ray Thornton's statements on risks were made to the September 10–11, 1981, meeting of the National Institutes of Health Recombinant Advisory Committee. (Copy provided by NIH.)

GENETIC ENGINEERING: THE SCIENCE III

9

THE GENETIC CODE. We live and die by it, procreate, grow, mutate, change, or remain the same according to its dictates. This is no code of conduct, but something closer to a linguistic code: rules of expression and rules for literally translating the four-letter language encoded in the bases of the DNA chain into a quite different language of amino acid sequences that compose proteins. We and the rest of the living world are made with significant amounts of protein and virtually all our bodily chemistry is carried out by enzymes, which are just a particular kind of protein.

We're going to look more deeply into the process by which DNA *instructions* lead to living *things,* a process still being unraveled by molecular biologists. But first let's take another look at the highlights of that DNA language system.

A DNA molecule is composed of two long chains twined around one another, joined together by complementary bases like steps in a staircase. The order of the bases determines a set of instructions just as the order of letters in English does. The region containing the instructions for a single protein is called a gene. Messenger RNA (mRNA), a molecule related to DNA, is composed of similar bases, and such a molecule is built to copy a particular stretch of DNA using a cellular copying machine called RNA polymerase. The messenger then travels to the ribosomes, the cell's automated workbenches, which "read" the orders from mRNA's base sequence and translate them into a particular sequence of amino acids, the building blocks of protein. As we trace the procedure in detail, we'll follow all the way through the translation process using the genetic code.

Through the schematic microscope, the view in Figure 7 shows a DNA sequence of a hypothetical *E. coli* gene. In front of the gene region is the promoter, or control sequence. Next to the promoter, pictured below the DNA sequence, is the mRNA copy that has been formed, and below that is the sequence of amino acids that will be called for by that messenger RNA. Notice that the two DNA strands have different names—template and coding—and that makes sense given the principle of complementarity. The coding strand has the *same* sequence as the messenger RNA, except for the substitution of the base uracil for thymine in mRNA. The other strand is called the template because the RNA polymerase copying machine actually works by *reading* the template strand, then assembling its complements. That forms a messenger RNA identical to the coding strand of DNA.

The figure shows several special sequences that must be read to produce the mRNA copy: the promoters and the start and stop sequences. The promoters are sites recognized by RNA polymerase. The sequences at these sites bind to the RNA polymerase, allowing it to "sit down" on the DNA to begin copying (transcribing) mRNA farther downstream (to the right in the figure) on the molecule. Most promoter site sequences in *E. coli* are similar to those shown—similar, but not necessarily identical to one another, another interesting property. For example, this sentence isn't right: "The qik bro fx jumpeed ov thr lazi dig." But despite the many typos, it contains enough information for you to recognize it. Similarly, many promoters are alike enough to be recognized by RNA polymerase—although here, no one of them is "correct English,"

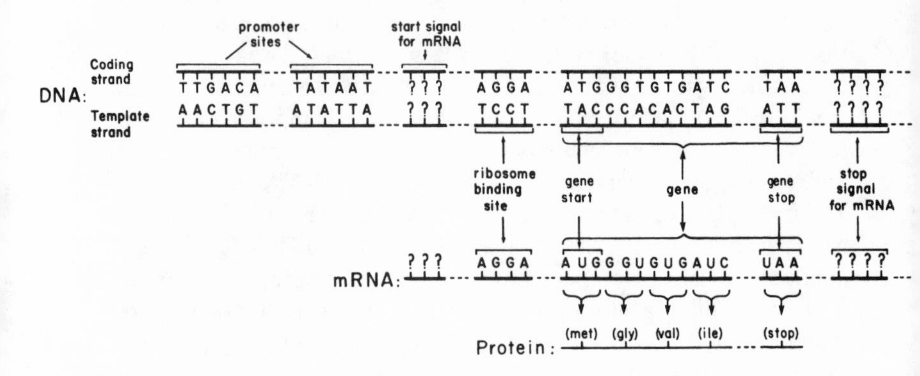

 Figure 7. Reading DNA language.

while the others contain errors. All are near variations of one another. If you were handed a piece of DNA, in most cases you, too, would be able to spot the promoter. In *E. coli,* the promoters will contain sequences identical to or similar to those pictured here. For example, they all contain a version of the "Tata Box," that is, the sequence TATA—that's how investigators recognize promoters. Recognizing promoters is important because they indicate a gene nearby downstream, and you would be able to find it. Furthermore, you would even be able to say what protein would be made on the ribosomes from that gene, once you've learned to speak in genetic code.

The sequence of the bases in the promoter region determines how strongly RNA polymerase binds. Thus, if the promoters are strong ones, at any given time more RNA polymerase will be making messenger RNA. Genetic engineers are beginning to learn a lot about which promoter sequences are strong. In nature, cells often have strong promoters in front of genes for proteins they need in large quantity, and weak promoters in front of genes that are rarely expressed—although this is not the only method of controlling protein production.

The remaining two "signal sequences" are the "start" and "stop" signals. They tell the RNA polymerase to start copying just before the beginning of the gene as it moves along the DNA and to stop just after the end of the gene. In Figure 7, the start signal is shown by question marks because there are several variations of such signals. Often the sequence . . .CAT. . . is found near the start, but often not. Start signals for many organisms have been described, but we know little of the stop signals; there are some stop regions rich in G and C bases. We do know that reasonably definite stop signals exist. Thus, the mRNA copy of the gene contains the base sequence coding for the gene plus some bases on either end of the gene.

Now let's study the mRNA copy in Figure 7 more closely. You will notice that the sequence . . .AGGA. . . is the ribosome binding site, and, with the uracil replacement of thymine, we can guess that the mRNA will bind at the ribosome's complementary sequence . . .UCCU. . . so that protein expression can begin. Other sequences are associated at one time or another with ribosome binding, but AGGA is the most frequent.

The strength of the binding site is important, and genetic engineers are trying to isolate other strong binding sequences, because if the bond is weak, the mRNA can fall off the ribosome before the protein is started. Again, having located the promoter sequence, you could find the ribosome binding site easily by looking downstream for the AGGA sequence; you might miss it, though, if you saw only an AGG sequence or another simple variation that, like the promoter, is close enough to be readable.

Now we come to the most important sequences of the mRNA, those which code for the various amino acids that will be assembled into a protein: the gene sequences. Since mRNA is composed of a four-base alphabet and proteins are composed of various combinations of these twenty amino acids, we need to learn to translate DNA language into protein language. Here it comes.

A Bioblitz Course in Genetic Code

Given a sequence of mRNA bases that has been transcribed from the DNA gene, what amino acid will be ordered up? The answer turns out to be simple, using the mRNA-protein dictionary that follows. This dictionary is literally what molecular biologists mean by the genetic code. Learn to use the dictionary and you can read any gene for the protein it will express. First, in the genetic code, all "words" are composed of three base letters. That means that three mRNA—or DNA—language bases (also known as a codon) determine one amino acid. For example, anytime you see the sequence GGU on mRNA, one molecule of the amino acid glycine will be incorporated into the protein being assembled on the ribosome, like another bead on a string. If the mRNA sequence reads GUG, the amino acid ordered is valine.

The genetic code dictionary for the twenty amino acids is shown in its entirety in Table 2, and here are a few examples in reading it. Consider the mRNA triplet AUG. The table says if A is in the first position, move to the third row down under the label First Position of the table. The second position in our mRNA triplet is U, located at the first column in the table's Second Position listings. Now you should be looking at the listings:

Ile
Ile

Ile
Met

The third position of our triplet is G, so from the instructions at the right-hand side of the table, you can see that the amino acid coded for is methionine (Met). Work out a few more for familiarization and find these answers: UUC codes for Phe (phenylalanine), GUU for Val (valine), GUC also for Val, AGC for Ser (serine), and AGA for Arg (arginine). How come *both* the triplets GUU and GUC will yield valine? We'll explain that later, along with the headaches that type of duplication might cause genetic engineers.

All that remains in learning to read DNA language is to learn the stop and start signals encoded in mRNA for making protein. It turns out that AUG is almost always the start signal for the ribosome. That is, when the ribosomal machinery hits AUG, it picks up a methionine amino acid and begins the string of amino acid beads. Therefore, almost all *E. coli* proteins begin with a

Table 2. The Genetic Code

	SECOND POSITION				
First Position	**U**	**C**	**A**	**G**	**Third Position**
U	Phe	Ser	Tyr	Cys	U
	Phe	Ser	Tyr	Cys	C
	Leu	Ser	Term	Term	A
	Leu	Ser	Term	Trp	G
C	Leu	Pro	His	Arg	U
	Leu	Pro	His	Arg	C
	Leu	Pro	Gln	Arg	A
	Leu	Pro	Gln	Arg	G
A	Ile	Thr	Asn	Ser	U
	Ile	Thr	Asn	Ser	C
	Ile	Thr	Lys	Arg	A
	Met	Thr	Lys	Arg	G
G	Val	Ala	Asp	Gly	U
	Val	Ala	Asp	Gly	C
	Val	Ala	Glu	Gly	A
	Val	Ala	Glu	Gly	G

methionine. The machinery continues reading bases in sets of three until it reaches a stop signal, coded for by the three triplets UAA, UAG, or UGA—shown as "term" on the table, for terminate.

Here's how all this translates for genetic engineers. They want to transfer the gene for human interferon into *E. coli* so the bacterium will make vast quantities of the rare human protein. They have tentatively identified a cloned colony containing the interferon gene. To make sure the desired gene is there, they will determine the base sequence of the area of the DNA around the gene using the experimental technique of DNA sequencing, and will thus know the corresponding mRNA sequence. After locating the start signal, they can translate that mRNA sequence into an amino acid sequence to see whether it is the one for interferon. Moreover, by sequencing the DNA *up*stream from the gene region, they can be sure that the ribosome binding site and promoter are at the best distance from the gene to achieve maximum mRNA production and maximum ribosome binding strength.

The actual translation process at the ribosome is tortuously complicated, but luckily we can understand some of it fairly quickly, remembering the principle of complementarity in binding. The mRNA is moved along the ribosome so amino acids can be tacked onto the growing chain, but the *actual* translation from RNA to protein language is carried out not by the ribosome but by other adapting molecules, another species of that varied RNA molecule called transfer RNA—tRNA. It is the tRNA molecules that actually carry the individual amino acids to the ribosome. Thus, the primary function of the ribosome workbench is to bind and orient both mRNA and tRNA so that the protein can be formed.

There are many varieties of tRNA already "made up," unlike the mRNA, which is assembled as needed. Each variety of tRNA recognizes two things: one amino acid and one mRNA base triplet. When found on the scene of protein formation (Fig. 8a), the clover-shaped tRNA already has the appropriate amino acid bound to it, just as with a little yard engine that is specific to one and only one kind of freight car. Figure 8a shows two of the many types of tRNA—one specific for phenylalanine and one to alanine. The "real-life" tRNA chain folds back on itself in a three-dimensional structure not unlike the clover leaf shown. While this folded tRNA structure is complicated, we need only look at two features. First, the anticodon region, so named because it is complementary to the codon region of mRNA. In phenylalanine tRNA, that region consists of an AAG sequence, at the tip of the middle clover leaf (the tRNA molecule has about seventy-five to eighty other bases not shown). Naturally, that makes it complementary to an mRNA's UUC sequence, and that is where it will interact with the mRNA. Similarly, the CGG anticodon on the alanine tRNA complements the GCC codon on mRNA, and this tRNA carries alanine bound to itself. Check these codes out in the table.

In Figure 8b we are witnessing the assembly of a protein—a glob, which is what most proteins look like—with another amino acid, phenylalanine, being added on. First, the mRNA lines up on the ribosome so that the codon for phenylalanine (UUC) is exposed to the tRNA anticodon. Then, the tRNA lines up by binding to the codon through the principle of complementarity, so that the amino acid it is towing is in proper position to be hooked onto the growing amino acid chain by an enzyme. In Figure 8c the mRNA has moved down one notch (three bases) on the ribosome, so that its triplet GCC, which orders up alanine, can be read. The tRNA for alanine positions it in just the right place for tacking on. In this case, the alanine is tacked on to the phenylalanine amino acid.

This process is as marvelously precise as any industrial robot's reading and carrying out instructions from punch tape.

We mentioned the headaches caused genetic engineers by the fact that several messenger triplets can order up the same amino acid. The problem: Not all organisms have all the transfer RNAs. Here is what happens: say the mRNA from an animal cell sometimes uses the code CUU for leucine, and other times uses UUA for it. But the bacterium in which the genetic engineer wants to insert a gene from an animal cell *only* uses the triplet UUA for leucine. Therefore, it might not have a leucine tRNA that will recognize the messenger's CUU, which means it can't synthesize any protein containing a CUU triplet. In effect, we're saying the bacterium doesn't have a complete, unabridged translation dictionary. The multiple triplets coding for one amino acid can be thought of as synonyms, and in this case our humble bacterium recognizes "car" but has no idea what "automobile" means. Nevertheless, once the lack is uncovered, it may also be possible to

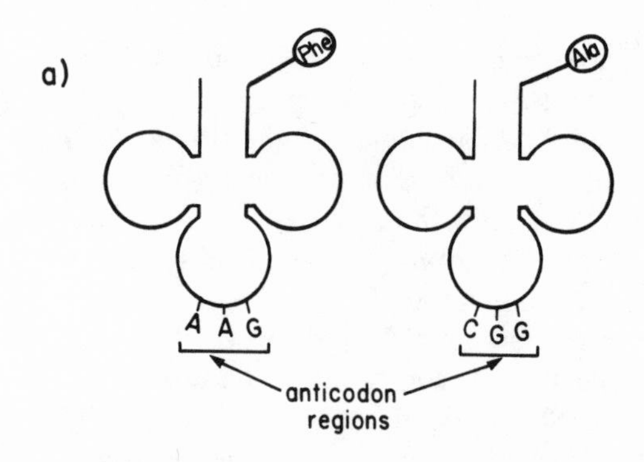

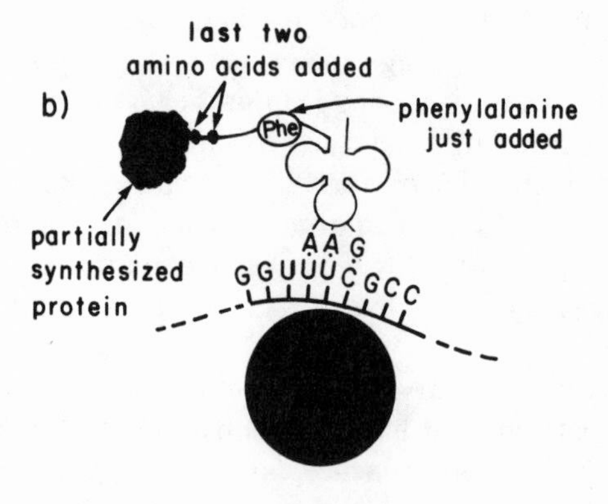

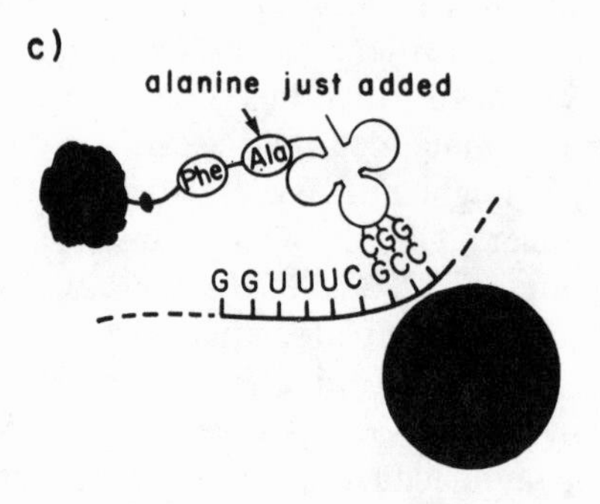

Figure 8. The process of translation.

engineer into the bacterium the instructions for making the CUU-recognizing tRNA—that is, for making its dictionary more complete.

At this point, you know enough genetic engineering and its basic science to understand much of what is done in the field. But if you have a deeper interest, read on.

dvanced Genetic Engineering

)NA Cloning, Enzymes, Secondary Metabolites, and Scale-up

Higher organisms, such as animals and plants, have a more complex gene structure than bacteria, and that means different procedures have to be used to clone these genes into bacteria. The only technique now available to get the more complex genes is copy-DNA (cDNA) cloning. Shotgunning doesn't work because of a strange, unexplained trait of many of the genes of higher organisms: they contain "introns" and "exons" (no relation to gas stations or freeway ramps), and bacteria do not. Imagine you're reading a book and after thirty pages you come to pages of gibberish, scrambled letters that make no sense, then after a page or two the story continues; after a while, more gibberish, then the story goes on again. Exasperating for you, and exasperating as well for genetic engineers: reading along the DNA sequence of a gene, they suddenly encounter gibberish, then the gene goes on. These "nonsense" DNA regions are easy enough to recognize, because when the gene is translated into protein, the gibberish regions don't wind up reflected in the protein sequence.

Scientists call those gene regions that are actually expressed as protein *exons,* presumably for expressed regions, and the nonprotein-coding gibberish sequences *introns,* or intervening regions. Not only bacteria but many yeast genes lack introns. That is, all their DNA bases between the start and stop signals are expressed as protein.

Why do some organisms have introns within their gene sequences, and others not? No one is certain, but Keith Backman, a pioneer in overproduction of gene expression, explained a current theory, one that suggests that introns serve a function in sorting out genes that speeds up evolution. Generally, proteins are organized into functional domains. That is, one twisted bit of surface on a globular protein might be its point of attachment to a surface; another knob of different shape and charge might be its "business end" if it is an enzyme. If you picture evolution as a speeded-up movie, with events that take a million years going by in a few minutes, you can see that once a particular shape has evolved to accomplish a given task efficiently, it would be useful to keep the code for it together, and to keep in separate clusters other useful bits of information. Then we would have a form of evolution involving the sorting of these bits in various combinations—just like shuffling a deck of cards. According to this view, the value of intron regions is that they promote sorting during evolution, and actually allow evolution of entire kingdoms to proceed faster.

Consider the old saw that a monkey pecking at a typewriter, given eternity to combine letters and spaces randomly, would at some point write all the works of Shakespeare. If we changed the rules so that once a word was formed, spaces were set around it so that it could be repeated, we would reduce the amount of time this task would take manyfold, because at some far earlier point in the random sorting we would have all the words of English assembled and need only a few more eons to randomly sort them into the right order.

In this view, bacteria and other microorganisms had their introns edited out at some evolutionary turn; staying "micro" was their evolutionary strategy, so their genes were stripped of all but essential sequences. Indeed, bacteria mutate constantly to create numberless *variants,* but considering that the animal kingdom alone has diversified into everything from mosquitos to electric eels to humans, it has evolved a great deal more than have the microorganisms. A fascinating hypothesis, though still just that.

What happens to the introns between transcription and translation is known. In transcription, the introns are copied into the messenger RNA along with the expressed gene regions, then they are spliced out of the mRNA before it reaches the ribosomes, as shown in Figure 9, where expression is sketched for both bacteria and intron-containing organisms. The protein-coding DNA (i.e., the gene or exon) is denoted by thick lines, the noncoding DNA (i.e., introns) by thin lines, and hash marks show the gene start and stop signals. On the right side of the figure, E marks the exons,

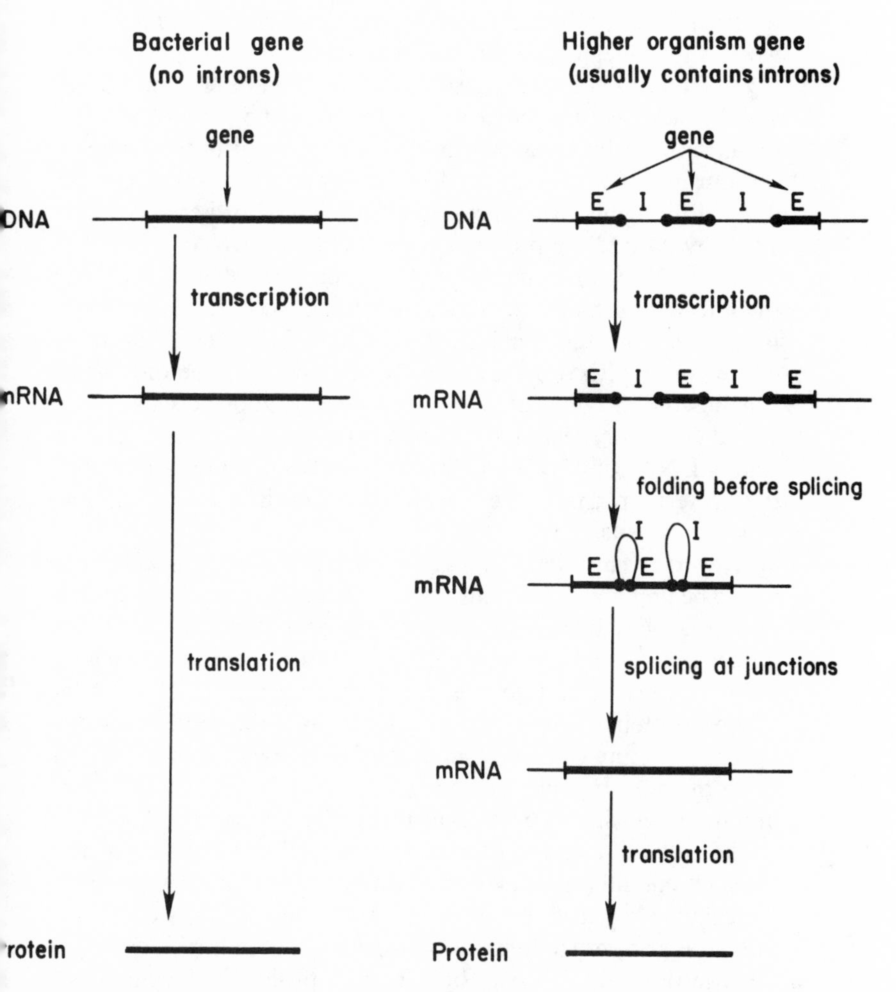

Figure 9. Comparison between the processing of bacterial mRNA and higher organism mRNA.

I the introns. Notice how the much more complicated higher organism's mRNA first is folded up inside particles called ribonucleoprotein particles so that the splice junctions (the filled circles in the figure) are aligned with each other. Enzymes now cut at the splice junctions, the introns are discarded—much as a film editor might cut certain scenes—and the remaining exons are spliced together. Net result: messenger RNA minus introns, after a lot of work whose meaning still is not clear to us.

Whether or not we understand the existence of nonsense DNA, we are faced with it when we try to splice the gene of a higher organism into bacteria, which have no mechanism for removing the introns. If we shotgunned the genes as described in Chapter 4, we'd find gibberish expressed in our cloned bacterial colonies. The way around the problem is difficult: copy DNA.

We start by obtaining the mRNA from the cell *after* it has had the introns taken out, then make a DNA copy of it in a backward formation from the usual DNA-to-mRNA direction of synthesis carried out by RNA polymerase. As we mentioned before, nature supplies the enzyme for this backward formation, discovered by David Baltimore, the aptly named reverse transcriptase. The resulting DNA copy is called cDNA. The cDNA is then spliced into the bacterial plasmid, the plasmid is placed into *E. coli,* and we proceed as usual.

Reverse transcriptase makes it possible to do what otherwise would be impossible. Cloning some interferons or any large animal-protein gene in bacteria could not have been done without it. But it does not make these engineering feats easy; the procedure is far more tedious than it sounds. First, finding the mRNA for the desired protein is sometimes very difficult. If the protein is a rare one, then at any given moment very few if any of its mRNAs will be in the cell. If genetic engineers are very lucky, they find conditions under which the normally rare protein is made in abundance so the cell will contain a lot of the right mRNA. But the search for such conditions, as with the search for the mRNA needle in the cellular haystack, can take years. Making a cDNA copy from mRNA is also very tedious, as is preparing (or "tailing") the cDNA for insertion into *E. coli.* Right now, a project involving cDNA cloning—and that means many of those we've talked about—could take from one to several years or might not be feasible at all yet. This helps to explain why achieving expression of interferon was expected to take so long, why the early success was truly such an achievement—and why we can't assume other feats will go as smoothly.

Improvements in procedure over the next few years should make cDNA cloning somewhat easier. Nevertheless, these uncertainties are serious ones for the research director, potential investor, or consumer wondering whether a new miracle product will soon be available.

econdary Metabolites

oking for the Fast Lane

Some of the greatest efforts and most revolutionary developments in genetic engineering's next phase will probably be in the improved genefacture of those vital products with unwieldy classification, "secondary metabolites" like alcohol and antibiotics, as well as primary metabolites like vitamins. But secondary metabolites, unlike many proteins, usually result from a long and quite delicately balanced series of reactions in the cell, so producing them presents special challenges. Biochemists call such progressive sequences of chemical reactions within a cell "metabolic pathways"; the movement on these pathways isn't necessarily from place to place, but from chemical to chemical in a chain of reactions.

Figure 10 shows a fictional metabolic pathway, and letters A through I represent different chemicals produced in the reaction pathway. For example, if we were making alcohol in yeast, A would be glucose, the sugar/food source, and F would represent the secondary metabolite, alcohol. H and I might be chemical products necessary for the cell to function, and to make both of them requires all the intervening reactions: A to B to C to D to E to I, and the product H is made along the way. Remember that in the cell these reactions are catalyzed by enzymes, and this particular metabolic pathway involves nine of them—e_1 through e_9. Arrows show which way the reactions are moving.

If this sketch begins to look like the traffic pattern along city streets, it's because these reactions behave much like flowing traffic, too. Think of A and B and the other capital letters as street intersections. Point F is the football stadium on the day of the big grudge match, and our problem is to get as many cars there just as quickly as possible. But there's always a bottleneck: the street between C and D has cars parked on both sides, so traffic narrows to one lane. That means that no matter how fast cars get from A to B and from B to C and so on, they will arrive at the parking

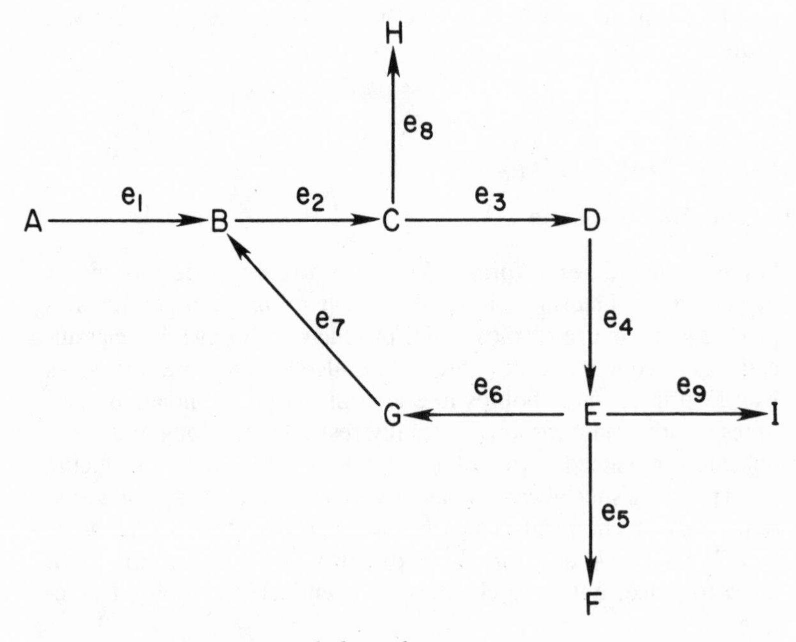

Figure 10. A fictitious metabolic pathway.

lot only at the slow crawl from which they get from C to D. This bottleneck principle actually applies in analyzing both traffic patterns and metabolic pathways.

Now look at Figure 10 as a series of chemicals and the reactions between them. The rates of reaction largely depend on which enzymes do the catalyzing: if e_3, which converts chemical C to chemical D, works at a slow crawl, then no matter how fast all the other enzymes along the pathway work to form the products B, C, E, F, getting to product F will take at least as much time as the slowest step, the formation of product D. Here is the bottleneck again, and having found it, we can see the direction in which solutions may lie.

The genetic engineer might find in another organism a better enzyme to change C to D and engineer it and its control sequence into the organism that makes our product. The new, fast-working enzymes will remove the bottleneck and the new rate of production will be on the scale of the *next* slowest reaction. Now we might get fancy, finding better enzymes for each step, putting strong promoters in front of their genes to get more of the new enzymes, and vastly increase the rate of production of F. This is precisely what genetic engineers are trying to do with alcohol production. Alcohol's metabolic pathways and enzymes are well known, so we ought to be able to beef up enzyme concentrations in the slowest steps.

Another trick to get more product F: destroy the functions of enzymes e_8 and e_6, so that everything is funneled straight to F. Means to destroy enzyme function were developed long ago by randomly changing bases of their genes into different bases. This is the chemical meaning of mutation. Some altered base sequences might occur in the control region of the gene itself. Mutations in the gene might lead to an enzyme with a reduced rate of catalysis or none at all, meaning the cell either can't carry out that reaction or has to do it by some other pathway. In Figure 10, if by mutation we can stop enzyme e_8 from working, then everything that funnels into C must go on to D; none will be siphoned off into H. If we can also knock out e_6 and e_9 by mutation, everything that starts out as A must end up as the desired product, F.

Superficially this seems like the "best route" from A to F, but it may not be. Chemicals H and I might be necessary to the cell's continued life. Knock them out and we may kill a valuable colony, and to paraphrase the old detective-story line, dead cells don't metabolize. (If we wanted a legal definition of death for a single-cell organism, it might well be cessation of metabolism.) That means the genetic engineer is always walking a tightrope when tinkering with metabolic pathways in cells. Even if an alteration does not kill the cell, it might slow its growth, and that in turn could wipe out the benefits of speeding up the desired chemical reaction. Another disadvantage of this technique is that, when it is used, one must search through thousands of clone colonies to find those that have the right kind of mutations, and in trying to beef up production of other products, the right one might never be found—it might not even exist, due to the randomness of mutation.

On the success side of the ledger, the amino acid lysine has been produced in large amounts without unwanted side products using the random mutation technique.

Genetic engineering may revolutionize this old procedure, too. The scientist knows beforehand what enzymes are to be beefed up, and can therefore *direct* changes in the metabolic pathways by finding the right enzyme and engineering it into the desired organism. The genetic engineering of metabolic pathways adds a new bag of tricks to the production of important secondary metabolites. Eventually, whole new pathways or portions of these pathways might be engineered into an organism such as *E. coli,* which would then manufacture secondary metabolites foreign to it, just as the bacterium now manufactures foreign proteins. That could open up a whole new chemical genefacture industry.

Fermentation Scale-u

The Easy Part Gets Ha

No matter how successful the laboratory wizard might be, his or her discoveries have no commercial importance unless they can be scaled-up to whatever industrial levels are required for a particular product. In the case of some drugs used in tiny quantities, scale-up may represent only a routine problem. In the case of large-volume products, that same scale-up may involve difficulties beyond any met in the lab. So far, the products of genetic engineering have not been truly large-volume, and scale-up problems have been limited. As one genetic engineer put it, scaled-up projects almost never work the first time out—there are always *some* difficulties to be worked out. But no project yet has been abandoned because what worked in the laboratory could not be reproduced in volume. Economics of scale-up, of course, are another matter.

In the laboratory it is usually easy to grow microorganisms in flasks of one or two liters, but the volumes of giant industrial fermenters often run to the thousands of liters. Why should that pose problems, as long as the growing colony is surrounded by nutrient? First, there's the problem of sterilization. It's easy to sterilize a laboratory flask—you put nutrient in it, stick it in a small autoclave (which kills off bacteria with heat), turn on the steam, wait half an hour, and it's done. How about a 10,000-gallon fermenter full of nutrient? Enough heat to do the job is expensive to deliver, and even tiny leaks could pollute the nutrient broth with foreign organisms. One possible answer might be to develop "thermophilic" bacteria for large-scale jobs. They would thrive at a temperature that would kill intruding bacteria—unless, alas, the invaders also were among the rare bacteria to thrive at high temperatures.

Further, genetically engineered microbes often reproduce more slowly than their "wild" counterparts (civilization always has its price), because in order to make a foreign protein they are using some energy that might otherwise go into reproduction. It doesn't take much slowing to have radical results, yielding a fermenter full of a wild invading strain instead of the one painstakingly engineered.

Suppose you inoculate a fermenter with a billion cells, of which 10 percent, or 100 million, are contaminant wild cells. Suppose the natural cells double every twenty minutes, but the redesigned cells take forty minutes. In 200 minutes—less than four hours—the fermenter will contain about 102 billion natural cells but only 29 billion engineered cells. The lesson: even a tiny contamination by faster growing cells will invariably lead to crowding out of the slower growing ones. Even though the overwhelmed cells may still be making product, nutrient is expensive and most of it is going to nonproductive but very hungry contaminant cells.

A related problem could turn out to be even more serious in some cases, although it has not proved a major problem so far. There is a natural tendency for organisms containing plasmids—as most genetically engineered bugs do—to throw off the plasmids eventually. That occurs as the cell doubles, when one daughter cell winds up without any copies of the plasmid. If the engineered features don't grant the cell better ability to survive, then the cells without plasmid, though initially few in number, may grow very much faster. Since plasmid-shedding is natural, it could become a chronic headache in genefacture.

In the laboratory there is a fairly easy solution to the problem, but it may not translate to industrial scale. Along with the desired gene, the genetic engineer splices resistance to a particular antibiotic into a plasmid. Cells then are grown in a medium containing the antibiotic. *Now* those that throw off the plasmid die. It's a neat solution on a small scale, but antibiotics are too expensive to be used in large fermenters.

A possible solution lies in the use of molecular switches. We mentioned one type earlier: the repressors. Repressors keep genes "turned off" until their proteins are needed. For example, if a cell normally metabolizes (eats) glucose but suddenly is in a medium where another sugar is available that it can eat, it will want to turn off the gene that codes for a glucose-metabolizing enzyme and turn on the gene coding for one that metabolizes the new sugar.

In industry, we might use molecular switches to keep the gene for our product turned off until the fermenter is filled to the brim with cells. Then the gene can be switched on and the cells will begin making our product all at once. The cells will be virtually all the engineered variety: no energy will have been lost during growth since the desired protein product is not made during the growing phase, and a few contaminants or a few cells tossing off plasmids will multiply out to a continued small percentage.

Some types of switches involve repressors, which are protein molecules, and a special region of the DNA called the operator region. The principle behind such operator-repressor switches can be seen under the schematic microscope of Figure 11. Figure 11a and 11b shows transcription as it was discussed in Chapter 4. An RNA polymerase molecule (solid square) binds to a promoter sequence P. The RNA polymerase then moves downstream and transcribes messenger RNA, shown in Figure 11b. But there is a new element here, between the promoter and the start of gene sequence G, a DNA sequence labeled O, the operator component of the molecular switch. (Some operator sequences actually exist inside the promoter region.) In Figure 11c we return to the transcription process before the mRNA is made, where the polymerase is bound to the promoter, as in Figure 11a. But here the operator has bound to it the protein molecule called a repressor, shown as an ellipse with a pie-shaped wedge cut out of it.

How does a repressor repress transcription of mRNA? By physically blocking the RNA polymerase from binding to the promoter and/or by blocking the downstream movement of RNA polymerase, as shown in Figure 11d, thus halting protein production for that gene. If there is no sugar in the cell's medium that is metabolized by gene G's enzyme product, then there's no need to have a G switched on. Later the cell may find itself deep in that very sugar—shown in Figure 11e as little pie-wedges. Evolution has designed repressors so that they can bind to themselves "inducer" molecules, so named because they stimulate the production of enzymes governing their metabolism. Appropriately, the inducer

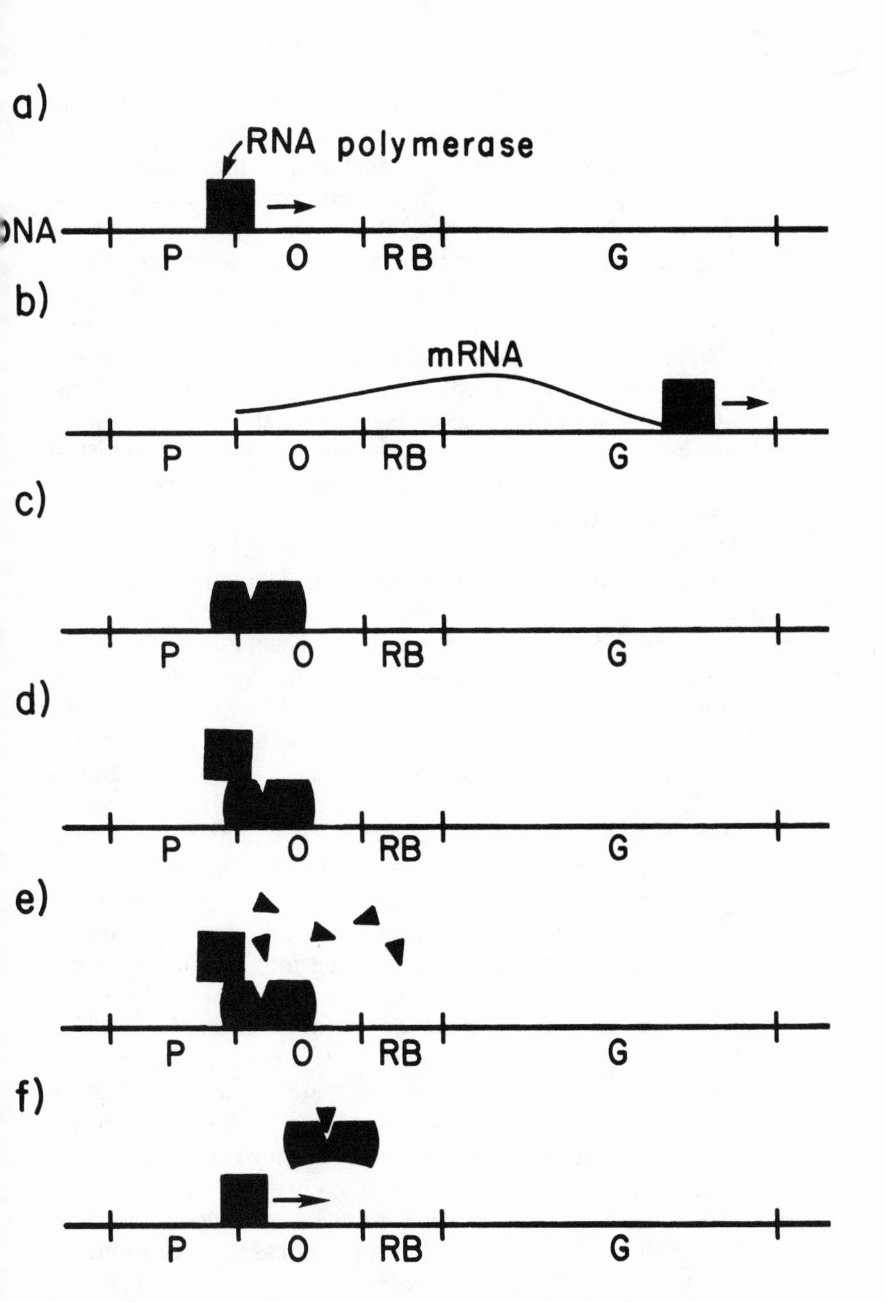

Figure 11. How repressor-operators act as molecular "switches" for transcription.

in this case is the very sugar that is now to be metabolized. Once bound to the inducer, the repressor can no longer bind to the operator region of the DNA. In other words, the inducer binds, the repressor is released from the operator region, and RNA polymerase moves downstream from the promoter to transcribe the gene (Figure 11f) that will metabolize the sugar (inducer).

Molecular switches can be used in genetic engineering by splicing operator sites along with the repressor gene into the appropriate places on the plasmid containing the product gene. That will keep the gene repressed while the colony is growing up at the fastest rate possible. Once the colony is grown, the genetic engineer simply dumps inducer into the broth and transcription begins. Genetic engineers are now trying to develop operator-repressor switches that use inexpensive inducers.

Molecular switches may also prove useful in cases where some secondary metabolites produced in desired large quantities may poison the producing bacteria. In this case, we could grow a giant colony of microorganisms, then use the molecular switch to obtain a large amount of end product by the time the colony died, whereupon we would begin growing a new colony. This would also enable us to overproduce end products beyond the same range for the bacteria; it would not matter that they used up so much energy that they could not produce a viable amount of their own proteins.

Returning to fermentation scale-up, contamination is not the only problem that must be solved. It can also be difficult to get air spread uniformly throughout the medium in a large fermenter.

E. coli and *B. subtilis* are aerobic bacteria, requiring oxygen in order to grow. Air might be bubbled into a giant fermenter containing the colony, but saturating the medium is difficult. Organisms such as yeast can grow with no oxygen (that is, anaerobically). Thus, genefacture of some products can be done in yeasts and other anaerobic organisms, where there are no aeration problems. It probably will take a few years to find the correct promoters and other expression-related features for most anaerobic organisms.

We have no doubt that even serious future scale-up problems will prove solvable, if not as easy to cope with as those encountered so far in industrial genetic engineering.

THE GENE AGE

10

WHAT BETTER PLACE to end than where we began, but speaking in the new language—at least, no longer needing to rely on so many metaphors and analogies to explain where we stand and where we are going. *A gene is nothing more than a region along a DNA molecule.* Now we understand that it does not stand out from the non-gene regions because it is a different color, or because it is composed of different bases; it is distinguishable because the sequences of the same four bases are set in particular order; they *read* differently.

Seeing the DNA molecule as we first did in Chapter 2, as two chains whose links twine together in a spiral double helix with its linear array of bases, we notice that at one point we pass the sequence TATAAT. Ah! The Tata Box. This must be a promoter region. Expect RNA polymerase to couple here and begin moving downstream, over the operator region where it will forge a messenger RNA's complementary initiation sequence and ribosome binding site. Passing the gene region, for example, we might notice the sequence TTC. When messenger RNA is transcribed along that sequence, the base uracil will replace the thymine (T) of the DNA, so the messenger will read UUC, and that tips us off to what will happen way down on the ribosomes. "There's a gene coding for the amino acid phenylalanine," we remark, noting from the genetic code table that uracil (U) in the first position followed by U in the second must yield either phenylalanine or leucine. With U or C (cytosine) in the third position, we always get phenylalanine.

If we could jump over to the ribosomes, a protein chemist might be able to tell us from the order of the amino acids, in turn, what protein was being constructed there—maybe a short-chain hormone, maybe an enzyme such as glucoamylase, maybe even one of the enzymes that would eventually come back to the DNA to continue transcription—such as the enzymatic protein RNA polymerase.

But for now, just as friends on a long, arduous trip might entertain one another by talking about the local sights and formations and confirming the witness of their eyes with a road map, we can talk about landmarks we might pass in some gene regions.

Medicine

The first human protein synthesized was somatostatin, a hormone produced in the hypothalamus that affects release of other hormones in the pituitary that are crucial to human fertility. But it was chosen for the first synthesis because it is so short—only fourteen amino acids make up its chain, and we can read them off easily:

Ala-Gly-Cys-Lys-Asn-Phe-Phe-Trp-Lys-Thr-Phe-Thr-Ser-Cys-STOP-STOP. We know that the corresponding gene region will be forty-two (3 × 14) base pairs long, and that it will begin with one of the codes for alanine: CGA, CGG, CGT, or CGC. It will end in that double stop that could be coded by one of the termination signals in the code: ATT, ATC, or ACT.

Researchers *synthesized* that string of nucleotides, an easy procedure if the protein to be made is short. They then provided the synthetic gene with sticky ends using one of Boyer's terminal transferase enzymes, allowing them to stick (or splice) the gene into an *E. coli* plasmid. The final DNA segment to be spliced was fifty-two base pairs long: forty-two coding for somatostatin, and ten providing sticky ends and a little of the information needed for expression of the gene. Scientists spliced this creation into a plasmid that also had added the control region and much of the gene region from an enzyme called β-galactosidase. When the β-galactosidase promoter began functioning to produce that enzyme, it produced the human protein as a tail on the end of it. The somatostatin then was chemically cleaved from its β-galactosidase carrier to yield the complete human hormone.

But obviously that kind of method works only with a short

chain, for tacking bases together is laborious. Human growth hormone is long by comparison—too long for synthesis of the gene. Human growth hormone genes can, however, be spliced into bacterial DNA. Figure 12 is a scanning electron micrograph of bacteria in their normal state. Figure 13, taken with a phase-contrast microscope, shows bacteria that have had human growth hormone genes inserted into their genetic array. The light white spots are the human growth hormone product that the bacteria have made along with their own normal proteins.

In practice, interferon was first synthesized by making a cDNA copy of the interferon messenger RNA from a strain of animal cells that overproduced interferon, thus making the usual needle-in-haystack search a little easier. The development of an mRNA overproducing strain and the screening for the proper clones were the difficult parts of this job.

With only a little new knowledge, it isn't hard to understand how these first, highly publicized genetic engineering feats were achieved.

Since those achievements a lot of progress has been made in finding strong promoters so many messenger copies would be made, and better ways for adjusting the promoters' distance from the desired gene to accomplish overproduction have been found. It all sounds similar to the tuning up of a complex piece of very precise machinery; it is.

Back in Chapter 1, we mentioned the possibility of creating completely safe vaccines against viruses. Here's how: viruses contain only a small amount of DNA, typically on the order of 10,000 base pairs—roughly the same as a bacterial plasmid's—split up into very few genes. Some of those genes code for manufacture of the virus's protective protein coat, on whose surface are those identifying markers that trigger antibody formation. On DNA molecules of this small size it is not difficult to find and clone the gene for one protein. (The 10,000 base pairs will be grouped into anywhere from five to twenty genes of somewhere around 1,000 base pairs each. Compare these 10,000 base pairs to one million base pairs in a bacterial DNA molecule or one billion in an animal's.)

To make such a vaccine, we just chop up the viral DNA with restriction enzymes, shotgun the viral genes into *E. coli,* and find the protein-coat gene. How? Inject the whole, living virus into laboratory animals and isolate antibodies to it. By means of a

Figure 12. Normal bacteria.

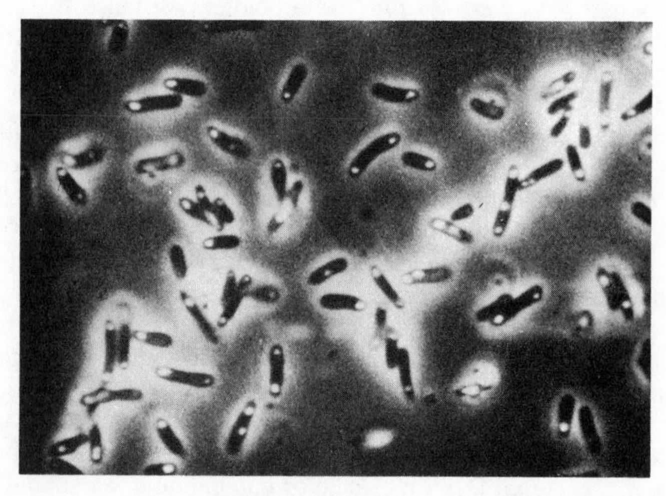

Figure 13. Bacteria genetically engineered to produce human growth hormone (white spots).

simple chemical reaction, the antibodies can be mixed with fluorescent or radioactive chemicals that will "tag" each antibody molecule. Now add the antibody carrying its beacon to each of the cloned *E. coli* colonies, and find the colony the antibody sticks to—it's got the coat protein, and the gene coding for it is in the bacterial plasmid. This recombinant-DNA-produced coat protein will now mimic a live- or killed-virus vaccine and will induce immune resistance to the whole live virus before infection.

The importance of such a recombinant DNA vaccine can be understood if you consider that if only one injection of the traditional killed vaccine out of 100,000 carries active virus, the results can be disastrous—possibly fatal to that one person per 100,000, but potentially so damaging to pharmaceutical companies that they have not made certain vaccines to avoid lawsuits.

In contrast, the recombinant-DNA vaccines made from genetically engineered virus-coat proteins are 100 percent safe, since they contain only a small, harmless part of the whole virus, enough to marshal our immune systems against future viral attacks. Many of these recombinant vaccines have been in the news recently. The Chiron Corporation of San Francisco is now clinically testing vaccines for hepatitis and malaria, and others are being developed for the wide variety of herpes viruses. Scientists hope that within the near future a recombinant vaccine against the AIDS virus may eliminate this rapidly multiplying killer before it becomes a true plague. Now you know the major strategy for developing these vaccines, why so many have come along in the past few years, and why they are truly safe. In the next few years, genetic engineering also may radically cut the price on existing vaccines. Live or attenuated virus for vaccines must be grown in very expensive cultures. *E. coli,* remember, likes sugar water with a few additives.

Let's go back to the story with which we opened *The Gene Age* concerning the phases of impact of new technology, and consider the effects of genetic engineering in medicine in light of it. We are already witnessing the replacement of existing vaccines and drugs with those created in bacteria, and here human insulin and the new-wave vaccines come to mind. Human growth hormone already is in use as a cure for dwarfism, but in recent months reports indicate that it and certain other human growth factors are being used successfully to help wounds heal quickly. Soon, according to a report in *Science,* human growth hormone may be tested as a means of slowing the aging process.[1] These latter developments ought to be considered carefully. The temptation will be great for all of us to use HGH to prevent our aging—but will sufficient tests be carried out to ascertain that there are no potential hazardous side effects?

Now that human gene therapy is on the horizon, how long will it be before the current ban on therapy in reproductive-cell genes is lifted to eliminate inherited scourges such as Tay-Sachs disease, sickle-cell anemia, Down's syndrome? Having come this distance through *The Gene Age,* we can see how this might be accomplished quite simply, in theory. These diseases all involve flawed genes—stretches of human DNA carrying garbled messages. We may not even need to snip out the erroneous gene; we'd probably only have to splice into the genome the correct gene. However difficult that may now be to carry out in practice, history tells us that solving technological problems requires only time. But beyond the curing of genetic disease, we will quickly encounter ethical questions of staggering proportions. Who will judge good from bad genes beyond the evolutionary law that suggests that bad genes promote the death of the organism while good genes enable it to live more successfully?

Industrial Chemicals: Manufacture

We've dealt with these chemicals already, but let's look over a couple in new detail, beginning with glucose isomerase, the enzyme that converts the less-sweet sugar found in grapes and blood to the supersweet fructose. Genefacturing this turned out to be a relatively quick and straightforward process, and it's a good illustration of how the techniques of recombinant DNA discussed here are put into practice.

We begin by finding the enzyme in a microorganism; since it occurs naturally at this level of life, the gene coding for it will contain no introns, those gibberish sequences that can make producing animal protein so difficult.

Now we select as host a mutant microorganism that *does not* produce glucose isomerase; that will be important when we try to find the gene, because we're going to do this job by shotgunning. In other words, we take an organism that *does* produce glucose isomerase and take out its DNA, cut up the DNA with restriction enzymes into many little gene-sized pieces (whose identities we don't yet know), splice all the pieces into plasmids, and then shotgun all these plasmid vectors into a colony of host microorganisms that *don't* produce this enzyme.

Next we follow the biologist's long-known cloning process: dilute the colony in solution so thinly that on the average each cell will have a centimeter of space between it and other cells, and grow up separate daughter-cell colonies from each original. One colony will be producing glucose isomerase (if all has gone right), for its original parent will have taken up the gene-bearing plasmid. This is exactly the procedure we studied under the schematic microscope as shotgunning. There is a simple color assay test to find which colony has the desired plasmid—a chemical combination that becomes colored when mixed with glucose isomerase.

The job's not done yet. We have plenty of genes available that code for glucose isomerase, but it's not likely they're tuned up just right. This could be where we separate winners from losers: Everyone attempting to make this commercially has found strong promoters, and promoter-to-gene distance had to be adjusted to achieve overproduction. Some experimenters are still trying to produce the enzyme in *B. subtilis* or another organism, so that the glucose isomerase can be secreted out of the producing colony and save the time and expense of purification.

Now let's move a step beyond in the genetic engineering process, applying our knowledge of enzyme pathways from Chapter 9. Looking back to Figure 10 (page 202), remember that the goal was to remove "roadblocks" so the more reactant chemical, A, can be converted into the desired product, F. We can now drop this hypothetical example and look at a real enzyme pathway engineered to yield vastly larger quantities of the amino acid phenylalanine than *E. coli* produce in nature. Phenylalanine is one of two key components of the artificial sweetener aspartame (NutraSweet). Like the hypothetical pathway of Figure 10, the pathway to phenylalanine production in *E. coli* has several slow steps, which created roadblocks. A major slow spot: a key enzyme in the pathway was inhibited by phenylalanine as it was produced. This so-called product or feedback inhibition is common among enzymes, one of nature's ways of regulating how much of a product is made. Thus, when too much phenylalanine is made by the *E. coli* cell, the phenylalanine will bind to all molecules of this key enzyme in the pathway, and that halts phenylalanine production at that step. The solution: locate a mutant enzyme from an entirely different microorganism that would *not* be inhibited by phenylalanine-product. Once such an organism was found, the key gene was found (using the methods we discussed earlier), and that gene was spliced into the host *E. coli* cell.

So much for the first roadblock. In the case of phenylalanine overproduction, several more roadblocks cropped up and had to be removed in a series of sophisticated steps. Then it was discovered that not enough of the reactant chemical—A in Figure 10—was available in *E. coli.* So genetic engineers had to back up still farther, to the steps in the cell that yielded reactant A to begin with, and they tinkered with those until they achieved vastly increased quantities of A to enter the phenylalanine pathway. The accomplishment required the close collaboration of several molecular biologists, enzyme chemists, and biochemical engineers. For their company, it confirmed the potential of metabolic-pathway engineering for chemical production.

An ever-increasing number of natural chemicals should be marketed over the next few years, produced via manipulated metabolic pathways, that will replace not only methods of direct chemical synthesis with their high energy requirements but production using nonengineered microorganisms. But remember, all enzymes come from living organisms, since by definition these proteins are natural catalysts. So it follows that only the products of living things can be produced by such enzymes. This excludes most of the useful industrial chemicals made in bulk, such as the molecular building blocks of plastics, which are not naturally occurring. Indeed, they are the creations of the brute-force methods of the petrochemical industry; no natural enzymes catalyze either their production or the production of their own intermediates.

Problem: Can we custom-make an enzyme to catalyze such reactions? An enzymatic protein is usually several hundred amino acids strung together on the ribosomes and then folded spontaneously so that each amino acid has a very critical position in the protein's surface or interior. That folding was shown schematically in Figure 5d (page 81). Such precise folding in order to carry out a catalytic function is the product of millennia of evolutionary experimentation. Obviously, one of the most vital segments of this final enzyme is the portion that binds to the molecule or molecules that are going to be reacted into the final product, and the shape of that portion is vital. The enzyme must bind to the reactant chemical or chemicals like a key fits a lock. This is shown schematically in Figure 4 (page 52), where the reactant A-B fits precisely into the pocket of the enzyme. Right now, except in a few simple cases, scientists cannot control the shape into which a protein folds, because they do not understand how the amino acids direct their own precise folding into a functional protein.

But recently, Richard Lerner of the University of California at San Diego and co-workers found a way to take advantage of nature's methods in order to make a *new* protein molecule with the catalytic properties of an enzyme. They were able to do so because of the remarkable ability of the mammalian immune system to custom-tailor antibodies to fit almost any invading antigen's shape. Antibodies are also proteins. The scientists injected a mouse with an antigen we'll consider to be just like our chemical A-B of Figure 4. The mouse's immune system automatically made antibodies with the right "pocket" to fit A-B. Then, using the techniques for making monoclonal antibodies, they prepared cells that would produce only the antibody that pockets A-B. But that's not quite the end of the road. Unfortunately, A-B will now key into the antibody's lock, but nothing will happen, because whatever further properties are required to turn A-B into the separated A and B product (denoted here by A + B) are missing. Those further properties were denoted in Figure 4 by the letter M, a group of amino acids residing in the pocket area, and we will assume now that the function of these amino acids is to stretch the A-B complex, to weaken its bonds and twist its shape so that as it spontaneously settles into a new configuration it will do so in the form A + B of our final product. Let's call that stretched and twisted shape A − + B; chemists would call this a transition state, and in many cases they can build molecules that mimic this transition state. Such a synthesized molecule, simply enough, is called a transition-state analog.

With such an analog in hand, we can make monoclonal antibodies to it by repeating the method used first time around. Inject the analog into a mouse, which will raise up antibodies to it, isolate the antibody and grow up clones of it—another monoclonal antibody. The new monoclonal antibody is added to the A-B mixture and indeed stretches and twists it into A − + B.

To date, Lerner's group has published a report on only one chemical reaction for which they have made monoclonal antibody enzymes.[2] In this model system, they were able to increase the rate of chemical reaction over a thousandfold above the noncatalyzed rate of reaction. An impressive number, for starters, but a rate increase of a millionfold would be much more promising. More significantly, three of twelve different monoclonals they grew up against the transition-state analog catalyzed the reaction, showing that the idea has real promise and should work on other chemical reactions.

The lab work required to make monoclonal antibodies is quite straightforward, but synthesizing transition-state analogs could be the real problem. Generally such syntheses are difficult, and generally, in order to work right, antibodies must be raised against the transition-state analogs. Though in its infancy, such enzyme catalysis of nonnatural products could be the cutting-edge biotechnology of the next century.

Industrial Pollutants: Cleanup

As mentioned earlier, much of the early promise for manufacturing industrial chemicals right now has been put on hold by the low price of oil. By contrast, uses of genetically engineered microorganisms for industrial cleanup is, in the words of one biotech executive, "going great guns."

Microorganisms exist naturally that can digest some industrial pollutants, but frequently they either can break down only a limited variety or do not break down the major contaminants efficiently enough to be of commercial value. The latter, for example, was the case with the first "bug" to be patented, the oil-spill-devouring microbe developed by Ananda Chakrabarty while he was with General Electric. Now at the University of Illinois, Chakrabarty has turned his attention to engineering organisms that can digest dioxin, the toxic component of Agent Orange that contaminates many military airfields around the world.

According to Bill Rosenberg, vice president of BioTechnica Ltd., the United Kingdom based subsidary of BioTechnica International, several firms including his own have had success developing mutant bacteria that will break down such categories of pollutants as cyanides, phenols, and what are known as "polynucleated aromatics." The strains so far have been developed using traditional forced mutation of the bacteria, seeking the best survivors to devour the chemicals, and for that reason environmental release should not bring the controversy it has with other, genetically engineered bacteria. In the absence of the offending pollutant, the new strains simply die back as the background population reasserts itself.

"There's the potential to create a 'cocktail' of our naturally selected bacteria that would go to work on a variety of pollutants you might find on a given site," Rosenberg said.

The idea of using natural bacteria to digest waste is not a new one. Municipal sewage waste now is spread on farm acreage in the Midwest, where soil bacteria break it down. The firms' strategy just carries this idea one step further, to break down the kinds of pollutants that might kill the now-abundant native bacteria. Rosenberg says problems with the EPA would not involve environmental release so much as how much of the pollutant is destroyed by the bacteria. "It will come down to 'How clean is clean,' " he said. "You'll never be able to achieve perfect, 100 percent cleanup, no matter what means you use. So how successful this is and how soon depends on what the standards are and how quickly we can meet them."

Agriculture in the Gene Age

Billions of dollars a year in crops are lost to diseases caused or carried by the same villains as are human diseases: viruses, bacteria, fungi, and insects. We discussed in Chapter 1 how disease-resistance had been cloned into tobacco via the toxin for *thuriengensis,* a protein that kills several insect species but is harmless to plants, humans, and other animals. Because the toxin is a protein, it was fairly simple to create the "new" tobacco plant by splicing the gene for the toxin into the tobacco plant along with the requisite promoter.

Even more advanced genetic engineering work is being done to give plants resistance to viruses. Of particular interest is a strategy called "antimessage," and we'll have to review a bit of how viruses do their dirty work to understand the antimessage strategy.

Remember that a virus is nothing more than a genetic blueprint for the virus (either in the gene regions of a DNA molecule or a related RNA molecule) surrounded by a protein capsule, whose function is mainly to protect it from the outside environment, and a small number of other proteins that enable the virus to invade specific plant or animal cells in order to be reproduced. A virus is so simple that it does not even contain digestive enzymes for energy and it may have no enzyme-tools to build the protein or nucleic acid components it must have to survive. For that reason, viruses inhabit a gray area between the nonliving and living worlds, lacking most of the functions we consider part of the definition of life, yet able to reproduce. Or more accurately stated in many cases, able to be reproduced, once they've hijacked their hosts' cellular reproductive machinery. The evil Doctor K of Chapter 2 picked up his tricks from viruses, learning to insert the instructions for a fancy sportscar into Mister L's robot workbench. As nature's genetic engineers, when viruses invade cells they need only the little gear required to divert the transcription/translation machinery to the making of viral reproductive proteins. One host cell might make tens to hundreds of new viruses as infection proceeds. This diversion of cells' machinery from synthesizing their own vital proteins leaves us with illness and debilitation and prevents crop plants from flourishing or kills them. In some cases invading viruses only temporarily disable their host cells, but in others the invaded cells die once they have reproduced many copies of their virus.

A typical virus might have a blueprint of ten thousand base pairs—about ten genes—compared to five million base pairs and thousands of genes in a bacterium and five billion base pairs and around one hundred thousand genes for an animal. Because a virus has so few genes, if genetic engineers can splice into a crop plant a means of interfering with transcription or translation of even one of these genes, they should be able to prevent viral reproduction. That could significantly increase crop yield and in some cases prevent loss of an entire crop.

The antimessage concept involves interfering with translation of viral genes' messenger RNA inside the host cell. When the virus takes over the cell machinery, its DNA is transcribed into messenger RNA; this is a single-stranded molecule that will carry the code to the ribosomes for translation into viral protein. Figure 7 (page 190) showed this process. If the complementary RNA of this viral messenger were in the host's cellular soup, the two would bind into an RNA double helix, and that would prevent the message from being read out on the host's ribosomes, since the translation machinery reads only single-stranded messages. That is the basis of the "antimessage" idea. It should not prove difficult for genetic engineers to clone into a plant's cells a piece of synthesized DNA which, when transcribed, will yield the complement or antimessage of the viral gene. Consider the example of Figure 7. If the viral message reads AUGGGUGUGAUC . . . UAA, then the genetic engineer must build and insert into the plant a DNA segment that will transcribe into UACCCACACUAG . . . AUU. Because the similarly shaped uracil replaces thymine in DNA, the gene the engineer constructs will read TACCCACACTAG . . . ATT. Notice that the antimessage does not contain a ribosome-binding site. That is of critical importance; otherwise, the antimessage might be read out on the ribosomes, yielding a protein that would be harmful to the plant. If the concept works, the antimessage will simply float in the cell, awaiting the appearance of the viral messenger.

Whether this strategy will work widely in the simple way we've described it remains to be seen, but there is nothing here outside the standard practice of genetic engineering. Potential problems: the viral antimessage is made but is not stable in the plant cell or is broken down and digested by cellular enzymes; or making the viral antimessage puts an "energy tax" on the plants that stunts their growth even without viruses in the area. We can certainly imagine strategies to cope with these problems, but they might entail fairly sophisticated genetic engineering to work. Within a few years some crop plants should be engineered with antimessages, so we will know soon how simple an effective strategy it will be.

The antimessage strategy probably cannot be applied against higher disease-causing organisms, such as bacteria, fungi, and insects, since they carry out their own energy-yielding and reproductive functions inside their cellular membranes. Their transcription/ translation machinery thus is segregated from that of the plants they infect.

Couldn't such an antimessage strategy be carried out against viruses that attack animals and humans? Yes, though perhaps with greater difficulty. But in the case of humans, ethical considerations would outweigh scientific difficulty in deciding whether such techniques should be used.

Certainly, this last discussion has taken us to the technical limit we can reach in this book, but in simpler terms, consider these other quite lively possibilities for genetic engineering.

Nitrogen fixation is certainly the single most important long-term goal of agrigenetic engineering, but it should also be possible to clone into plant cells the genes offering protection against many devastating crop diseases, or genes that will yield proteins fatal to crop-eating insects, or proteins that will simply make the plant unappealing to the pests. If protection against herbicides can be spliced into the gene complement of food plants, then herbicides harmless to crops but fatal to weeds could be sprayed on farmland.

We also discussed human growth hormone at some length. Varieties of growth hormones for livestock are now being investigated, and a few (such as bovine growth hormone) will soon be available. Others that will speed the growth of food animals should be available within three years. Most of these growth hormones are fairly short molecules, so figuring out their corresponding gene sequence should not be difficult. If the hormones do not need post-transcriptional modifications, they can be grown in *E. coli.* If they do, their genes may be spliced into yeast or animal cells. The purified hormone would then be injected into animals.

This area, however, is controversial. Midwestern dairy farmers have argued against introduction of bovine growth hormone (BGH) because they fear it will put them out of business. Milk is already a surplus commodity, and the price is so low that few small dairy farmers can survive. They argue, probably correctly, that BGH will require them to buy more feed, driving their costs up further, and that the increased production will drive the price down, cutting back on their already slim margin. If so, however, that will mean that only large, corporate dairies can afford to use the hormone and absorb the higher feed costs. The price of milk would drop, and that might contribute to putting small farmers out of business.

But there would be no stopping the use of the hormone on those grounds. It would be yet another case in which agribusiness is contributing to the demise of small farming, and in a high-tech age, small farming unfortunately seems destined to go the way of the horse-drawn plow—into history.

But interestingly, Thomas Wagner of the University of Ohio (Chapter 1) sees genetic engineering developments that might turn the small farmer into an industrial or pharmaceutical producer. Wagner's main goal in developing so-called transgenic cattle is to create a breed that would grow only as fast as a normal cow but on *less* feed than current breeds require. If successful, that might eliminate one of the small farmer's hazards, the increased costs of added feed. But he envisions transgenic cattle leading to a far greater revolution in agriculture.

The grandson of an Ohio farmer, Wagner recalls learning while young that "my grandfather was successful because, although he sold cheaply, he didn't buy at all. He lived very simply, without many finished goods."

Wagner observes, "The modern farmer's problem is that he buys high-priced finished goods but still sells low-priced commodities. He has to buy clothes, appliances, all the trappings we all want, and those are finished goods with high value added. Even the fertilizer and farm machinery he must buy are high value-added items. Then he turns around and sells a commodity, which is always a low value-added item. If you can't change that, there's not much point in talking about saving the small farmer, because there are *not* a lot of people out there who are willing to give up the material things we enjoy and live a simple, basic life."

Food is priced low "because that is a political decision," Wagner says, and so the impetus to keep food prices low will continue indefinitely.

But why couldn't the same genetic engineering techniques that allow us to engineer bovine growth hormone overproduction into cows be applied to having the cows produce foreign proteins of commercial use to us? Obviously no simple task, it is one Wagner believes will one day be accomplished, turning animal farming into a cottage industry for pharmaceuticals and other specialty chemicals—the epitome of high-value-added products.

An example: casein is a milk protein that has many industrial uses—as a component of some glues, for example. Cows could be developed to produce far larger quantities of casein. Further, if casein is modified by adding molecules of the amino acid proline, it becomes an important gel-forming protein, and carrying out that modification in the cow instead of in the factory could make this altered milk a specialty item for the food processing industry rather than a commodity.

Such development is important not merely to agriculture, but to industry, Wagner noted, and he is in a position to see this first-hand. Ohio is the country's fifth leading agricultural state, but it is also a major industrial state, and one particularly hard hit by the loss of traditional American industry to foreign competition.

The Cutting Edge

The War Against Cancer

Generally we've discussed the commercial applications of recombinant DNA, and we think that understanding the complex concepts involved in all genetic engineering is made somewhat easier by looking at the relatively simple creatures used in industrial genefacture. Recombinant DNA technology, however, has had profound implications for basic research in molecular biology.

Some people think of "basic research" as referring to the arcane and incomprehensible, the work scientists do in their search for a "truth" whose meaning is hard for the rest of us to discern. Mention cancer, on the other hand, and everyone's interest quickens—yet research into the causes of cancer is just such basic research, and an area in which recombinant DNA techniques have brought scientists to major discoveries.

The most significant discoveries to date have been those that brought us far closer to understanding what cancer is. That may seem like a small advance, but in fact it is a giant leap forward. Until around 1980, little was known about what went wrong in someone who developed cancer, and this despite nearly half a century of committed research. The problem: the causes of cancer operate at a submicroscopic level, beyond even the theoretical range of the electron microscope; in the literal sense, then, we can never *see* what is going wrong as a cell becomes cancerous. Further, and more frustrating, in the broad sense we know what causes cancer—so many different agents have been implicated in causing cancer that learning about another one has become a joke to many people.

Certain animal viruses cause cancer—that much has been known for years. When these viruses infect an animal cell, they can cause it to divide and grow wildly. Such uncontrolled growth is the major distinguishing characteristic of cancer cells, threatening death to the entire organism from tumors or other factors associated with the abnormally growing cells.

But it has also been known for years that cancers can occur in the absence of viruses. Chemicals can cause cancers, whether those found in cigarette smoke or in some pollutants. And of course the oldest known direct cause of cancer is radiation, discovered only after most of the early researchers in X-rays died of cancer. For all those reasons, scientists were pessimistic that any single cause of cancer would ever be found, and if causes remained unknown, cures seemed beyond hope. Then, within the past few years by using recombinant DNA and other modern techniques, the causes of some viral and nonviral cancers have been traced to a single source—cancer genes, often called oncogenes. By such methods as DNA sequencing, the genes of at least one cancer virus have been studied extensively.

Harvey's sarcoma, a rodent cancer virus, has become something of a laboratory celebrity because of its importance in the hunt for the cause of cancer. Scientists made a remarkable discovery about Harvey's sarcoma: the virus contains an animal gene, and that same gene was one found in the DNA of normal, noncancerous animals—including humans. Further, the gene was identified as involved in regulating animal cell growth. In effect, this viral DNA was a naturally occurring example of recombined DNA. If invading viruses normally insert their DNA instructions into the host, this is an example of the opposite—the virus contains pieces of the host's genetic information. Why is it there? We still don't know, but this is the kind of discovery that makes a scientist's imagination soar. Cancer is abnormal cell growth. Suddenly we find that a cancer-causing virus contains a normal animal gene whose protein product is involved in *regulating* cell growth. Consider this scenario: perhaps a particular virus infects a human cell and takes over its protein-making machinery; now the animal growth-regulating protein from the virus is overproduced, causing the cell to grow out of control. It has been shown that cancers caused by Harvey's sarcoma virus indeed are associated with high levels of growth-regulating protein.

Other cancer viruses were examined and they, too, possess genes from normal human and other animal cells. We now presume that these oncogenes are also involved in the regulation of cell growth, though that still must be proved.

More speculative evidence also points to overproduction of normal animal-cell proteins as a cause of cancer. Long before oncogenes were found, researchers knew that chromosomal mutation was involved in the development of at least one form of human cancer, lymphoma. Many lymphoma patients had a chromosomal mutation called a translocation, which involves the movement of very large regions of DNA, including perhaps hundreds of genes, to a new location along the chromosomal DNA.[3] In this case, the amount of DNA moved is so large that the translocation can be seen in a microscope. That made its existence known well before the development of such exquisitely subtle techniques as direct DNA sequencing, now used to detect and study mutations.

Initially, such a translocation's association with cancer might seem wholly unrelated to oncogenes. But then investigators at both Harvard and Philadelphia's Wistar Institute discovered that the translocation occurs at the very point where the existence of the oncogene had already been postulated. More interestingly, the translocation placed this postulated oncogene right next to a *strong promoter,* used in the normal cell to produce antibodies in large amounts. Thus, the explanation for lymphoma cancer also seems to tie in to overproduction of a protein involved in growth regulation; it is not yet proved that the postulated oncogene in fact regulates growth.

The oncogenes responsible for Harvey's sarcoma and some other oncogenes show another interesting property. Though nearly identical to normal animal genes, they contain a few mutations, base changes that code for a different amino acid in the resulting protein. Researchers have found that a single mutation changing a glycine amino acid into other amino acids makes an oncogene more active. The mutated oncogene need not be overproduced to cause cancer but can do so in very small quantities.

A great many of the early discoveries concerning the actions of oncogenes in developing cancer were made in the M.I.T. laboratories of Robert Weinberg, whose investigators work in labs directly adjoining those of David Baltimore. Weinberg is a former postdoctoral student of Baltimore's, one to whom his former mentor refers as "my most successful postdoc." Weinberg set out to understand why cancer appeared to have so many disparate causes, asking himself if carcinogens might not have in common their action on a single gene or set of genes. It was in subsequently discovering many oncogenes closely related to but slightly mutated from normal genes that he established the view of cancer now current among most scientists, the so-called oncogene theory. The work in Weinberg's laboratory was the focus of our book *Target: Cancer.*

Most of the deadliest cancers appear not to be caused by viruses—in other words, the genetic damage is either inherited, caused by chemical carcinogens, or perhaps even caused by a spontaneous mutational error that turns a normal gene oncogenic. But, also within this watershed decade in cancer research, National Cancer Institute researcher Robert Gallo discovered the first known virus to directly cause cancer in humans, HTLV, or human T-cell lymphotropic virus. Shortly afterward, Gallo and co-workers discovered that a related virus, dubbed HTLV-III, is the cause of AIDS.

Most recently, two scientific teams, one led by Weinberg and one by Thaddeus P. Dryja, made a discovery of enormous import in understanding the biology of cancer. They found a gene *the absence of which* causes retinoblastoma, a form of eye cancer that usually appears in children. The correspondence is 100 percent: those who lack the gene, whether because of inheritance or damage after birth, always develop retinoblastoma.

Without recombinant DNA, none of this could be known. Remember that we are working at a level below possible sight, even when augmented by the electron microscope. What gene-splicing enables us to do is to manufacture protein molecules and their encoding genes in huge quantities, molecules that may be made in the cell only in one or a few copies. In effect, recombinant DNA used in this way becomes a chemical magnifying glass, because once a test tube full of a protein is expressed, you can analyze its exact composition. Similarly, huge quantities of particular genes can be analyzed by the technique known as gel electrophoresis and it can be determined to the letter what the composition of its bases is. This is why in Chapter 3 Sir Francis Crick referred to recombinant DNA and rapid gel sequencing as the major surprises that had thrown off his forecast of the future as he and Watson were elucidating the structure of DNA.

Recombinant DNA has become the most powerful tool used in the study of oncogenes, their DNA sequences, mutations, and relationships to one another. Typically, genes are present in an animal cell in at most a few copies. An average gene is about a thousand bases long. The whole human chromosome is more than a billion bases long. To study a single oncogene amid all the other DNA is almost impossible. Imagine that a researcher needs DNA weighing only a milligram—a grain of salt's worth—to carry out experiments on a given gene. Deriving it would require 1,000 grams of DNA, and that would be the entire amount of DNA found in a 220-pound human. Consider that cell cultures are grown in little round petri dishes (with a diameter of four inches, on which cells can grow only one-eighth of an inch deep) and the enormity of the problem becomes obvious.

Now the researcher takes a milligram or less of DNA and shotguns it into *E. coli* by methods we've already discussed. Next, the gene is located, and the clone containing the desired gene is grown up in any quantity desired and easily spliced out of the plasmid DNA on which it had been cloned. Thus, incredible *amplification* is provided by recombinant DNA techniques.

CAUSES ARE NOT CURES

Evidence is mounting rapidly that two closely related events—overproduction of normal cell growth-regulating proteins and mutation of smaller amounts of growth-regulating proteins—are key, initial causes of the complex path taken by a cell toward cancer.

The cause of cancer may soon be found, but we must not be overly optimistic about finding the cure. Nevertheless, this evidence suggests that only a limited number of events involved in growth regulation are responsible for tumor formation, and that is exciting because it offers hope that a cure for our most feared scourge *will* be found. Such a discovery would bring to a close one of the major quests in the history of science.

If recombinant DNA research has pointed toward the causes of cancer, it has also led to some remarkable therapeutic discoveries—the first major additions to the anticancer arsenal in many years. We've already referred to several by name: the interferons, interleukin-2, tumor necrosis factor. These are direct products of genetic engineering, and the latter offers a good example of where the "cutting edge" in recombinant DNA may lead us. Michael Kriegler, encountered initially at Fox Chase Cancer Institute, then working on TNF at Cetus, was one of the first investigators to make large quantities of the molecule, which is naturally found in cells in such tiny quantity that, like interferons, its function was barely known a few years ago. Kriegler noted that TNF causes death to cancer cells, but so far it also has unfortunate toxic effects on many normal cells, preventing its being used in large quantities to attack cancer.

But what is a virus except a protein coat engineered by nature to carry a genetic message into a very specific type of cell, there to reproduce the message? Kriegler believes that one day he may be able to engineer a virus so that it would attack only the tissue-type cells that have become cancerous; and into these cells he might deliver the genetic message to reproduce not only another such virus, but tumor necrosis factor. The TNF thus would always remain segregated from other types of cells, because once inside the body it would either be inside a cancerous cell doing its job or packaged in a viral coat protein. Whether this turns out to be the strategy that will be remembered for ending one of mankind's all-time scourges, such thinking shows the huge territory over which scientists' imaginations have been able to range using these powerful new techniques.

Conclusion

If there is one truth that emerges in this book, repeating over and over like a base theme in endless variation, it is the universality of this language that all life on earth holds in common, the language we've been describing, translating, trying to speak. And that suggests as well the great uniformity, congruity, or at least harmony to the notion of life that underlies the equally awesome differences between individual humans or between humans and other creatures, complex or microbial.

Why can humans metabolize the insulin of pigs and cows? Because the DNA chains of those homely animals along that particular gene-region are nearly identical to our own. But there is more similarity between all of "us" than that. An eye is an eye and a hair is a hair: there is more alike in the DNA sequences that code for these complex creations in different animals than there is dissimilar. The greater dissimilarity is between the eye and hair: How do identical DNA molecules "know" to order eye-building proteins at one place and hair-building proteins at another? The incredible process of cell differentiation, alluded to by several scientists in this book, still is one of the great unsolved biological mysteries. But not forever, not now that we've learned to speak the language, however crudely at this stage. At the level we've been traveling, of course, our *bodies* know the language fluently. Our "minds" may not, but our brain cells do. At that level, "we" speak the same language as the virus. If we did not, the virus could not infect us by taking over our own cells' mechanisms with such well-understood instructions. But then, we could also not have turned the tables, finally figuring out this great key to a boundless source of knowledge, energy, and tools that will change us forever.

NOTES

[1] *Science,* October 3, 1986, p. 23.

[2] "Catalytic Antibodies," by Alfonso Tramontano, Kim D. Janda, and Richard A. Lerner, in *Science,* December 19, 1986.

[3] "Translocation and Rearrangements of the *c-myc* Oncogene Locus in Human Undifferentiated B-Cell Lymphomas," *Science,* February 25, 1983. Copyright 1983 by the American Association for the Advancement of Science. Reprinted with permission.

INDEX

Pages containing definitions of terms or major sections on indexed subjects are indicated in **boldface**.

Sylvester, Edward J.
The gene age.